Irakli Titvinidze

Dynamical Mean-Field Theory Approach for Ultracold Atomic Gases

Irakli Titvinidze

Dynamical Mean-Field Theory Approach for Ultracold Atomic Gases

DMFT for Ultracold Atomic Gases

Südwestdeutscher Verlag für Hochschulschriften

Imprint

Any brand names and product names mentioned in this book are subject to trademark, brand or patent protection and are trademarks or registered trademarks of their respective holders. The use of brand names, product names, common names, trade names, product descriptions etc. even without a particular marking in this work is in no way to be construed to mean that such names may be regarded as unrestricted in respect of trademark and brand protection legislation and could thus be used by anyone.

Publisher:
Südwestdeutscher Verlag für Hochschulschriften
is a trademark of
Dodo Books Indian Ocean Ltd., member of the OmniScriptum S.R.L Publishing group
str. A.Russo 15, of. 61, Chisinau-2068, Republic of Moldova Europe
Printed at: see last page
ISBN: 978-3-8381-2094-2

Zugl. / Approved by: Goethe-Universität in Frankfurt am Main, Frankfurt am Main, Diss, 2009

Copyright © Irakli Titvinidze
Copyright © 2010 Dodo Books Indian Ocean Ltd., member of the OmniScriptum S.R.L Publishing group

Dedicated to the memory of my father

Contents

1 Introduction — 5

2 Ultracold atomic Physics — 9
  2.1 Short Overview — 9
  2.2 Cooling Atoms — 11
  2.3 Feshbach Resonances — 13
  2.4 BCS-BEC Crossover — 20
  2.5 Optical Lattices — 23
  2.6 The Hubbard Model — 27
    2.6.1 The Fermionic Hubbard Model — 27
    2.6.2 The Bose-Fermi Hubbard Model — 29

3 Method — 31
  3.1 Dynamical Mean-Field Theory (DMFT) — 31
  3.2 Generalized Dynamical Mean-Field Theory (GDMFT) — 38
    3.2.1 Method — 38
    3.2.2 Energy Calculations — 42
    3.2.3 Summary — 43
  3.3 Real-Space Dynamical Mean-Field Theory (R-DMFT) — 44
  3.4 Impurity Solvers — 47
    3.4.1 Exact Diagonalization (ED) — 47
    3.4.2 Numerical Renormalization Group (NRG) — 49

4 Mixtures of Fermions and Bosons in Optical Lattices — 55
  4.1 Mixtures of Spinless Fermions and Bosons in Optical Lattices — 58
    4.1.1 Half-Filled Mixture of Spinless Fermions and Bosons — 58
    4.1.2 3/2-Filled Bosons and Half-Filled Spinless Fermions — 62

	4.2	Mixtures of Hard-Core Bosons and Two-Component Fermions in an Optical Lattice	67
5	**Ultra-Cold Atoms in a Harmonic Trap**		**71**
	5.1	Ultra-Cold Fermions in a Harmonic Trap	72
		5.1.1 Balanced Mixture	75
		5.1.2 Imbalanced Mixture	78
	5.2	Mixtures of Spinless Fermions and Bosons in a Harmonic Trap	81
6	**Resonance Superfluidity in an Optical Lattice**		**85**
	6.1	Microscopic Model	86
	6.2	Method	88
	6.3	Results	93
7	**Summary**		**97**
8	**Zusammenfassung**		**101**
A	**Relation between Experimental and Hubbard Parameters**		**107**
B	**Derivation of the DMFT Effective Action**		**111**
C	**The Equation of Motion and Green's Functions**		**115**
D	**Derivation of the Self-Energy for a Bose-Fermi Mixture**		**117**
E	**Derivation of the Kinetic Energy**		**123**
F	**Iterative Diagonalization within NRG**		**127**
	Bibliography		**131**
	Acknowledgments		**145**

Chapter 1

Introduction

The impressive experimental progress in the field of ultracold atoms in the last decade has brought it to the forefront of research on strongly correlated quantum many-body systems. The possibility to confine and manipulate atoms in optical lattices created by standing waves of laser light gives the opportunity to realize some of the model Hamiltonians of condensed-matter physics, and this way shed light on notoriously difficult problems [1–3]. Going beyond that, also systems without clear analog in "conventional" condensed matter systems can be realized. In particular, cold atomic gases offer the possibility to realize mixtures of fermions and bosons [4–19]. This yields a very rich system, which at this moment is far from fully explored.

This system bears some analogy with the well-known two-component Fermi-Fermi mixture, but is in fact much richer. By replacing one of the fermionic components by bosons, one keeps the instability of half-filled fermions towards charge-density wave (CDW) ordering. For historical reasons we keep this terminology throughout this thesis, although the fermionic atoms under consideration do not carry a charge. At the same time the bosonic species can be superfluid, allowing for supersolid behavior, where diagonal CDW order coexists with off-diagonal superfluid long-range order. Several previous theoretical works have studied mixtures of fermions and bosons in an optical lattice [20–45].

Investigating a strongly correlated Bose-Fermi mixture in an optical lattice is a difficult problem, to which powerful numerical and analytical techniques have been applied. In one dimension this involved Bethe-Ansatz technique [25], bosonization [26, 28], Density Matrix Renormalization Group [32, 35], and quantum Monte Carlo (QMC) [23, 36–40]. In higher dimensions, however, non-perturbative calculations are sparse. In two dimensions Renormalization Group studies [24, 31] have been carried out. Although able to describe non-perturbative effects, this technique is limited to weak couplings. Another powerful technique that has been applied in two [29], and recently also three dimensions [33, 34] is to integrate out the fermions. In this way one generates a long-ranged, retarded interaction between the bosons, which means that the resulting bosonic problem is still hard to solve. Important progress has recently been made in mapping out the Mott-insulating lobes. A composite fermion approach [30] was used to qualitatively describe possible quantum phases of the Bose-Fermi mixture.

In this thesis we introduce and apply Generalized Dynamical Mean-Field Theory (GDMFT) to study this problem. This is a non-perturbative method which becomes exact in infinite dimensions and is a good approximation for three spatial dimensions (see section 3.2). The only small parameter is $1/z$, where z is the coordination number. For this reason, the method reliably describes the full range from weak to strong coupling. The advantage of this method is that in contrast to QMC calculations, this method works in high dimension and not only allows to map out the phase borders but also gives reliable results away from it (in contrast to the Refs. [29, 33, 34]). As we will show in section 3.2 GDMFT has a systematic derivation in contrast to the composite fermion approach.

We apply the GDMFT to a variety of cases. In particular, we study commensurate mixture of the spinless (spin-polarized) fermions and bosons, as well as a mixture of hardcore bosons and two-component fermions in an optical lattice. The reason why we chose commensurate filling is that in this case interesting phases, like the supersolid can occur, which break the translation symmetry. We also take into account the effect of the harmonic trap. For this purpose we develop Real Space Dynamical Mean-Field Theory (R-DMFT) (see section 3.3).

This thesis is structured as follows: In chapter 2 we present a short overview of the physics of ultracold atoms. We start with a short historical outlook (section 2.1). Then we briefly review the cooling methods (section 2.1), Feshbach resonance (section 2.3), BCS-BEC crossover (section 2.4), optical lattices (section 2.5) and in the end of the chapter we derive the model Hamiltonians that will be dealt with the rest of the thesis (section 2.6).

In chapter 3 we describe the methods used in this thesis. In particular, dynamical mean field theory (section 3.1), GDMFT (section 3.2), R-DMFT (section 3.3). In this chapter we also consider impurity solvers which we use during our calculation: exact diagonalization (section 3.4.1) and numerical renormalization theory (section 3.4.2).

In chapter 4 we study the mixture of the spinless fermions and bosons for commensurate fillings. In particular, in section 4.1.1 we study this mixture when both of them are half-filled, while in the section 4.1.2 we study the case when the fermions are again half-filled while the filling of the bosons is 3/2. In this chapter we also study a mixture of hard-core bosons and two-component fermions (section 4.2).

In chapter 5 we study the effect of the harmonic trap. First we study a repulsively interacting two component Fermi gas in a harmonic trap. We investigate the stability of the antiferromagnetic order against the presence of the harmonic potential (section 5.1). Later on we consider a mixture of spinless fermions and bosons and investigate the stability of the supersolid in the presence of the trap (section 5.2).

In chapter 6 we study an ultracold atomic gas of fermionic atoms and bosonic molecules close to a Feshbach resonance. We consider the process when due to the Feshbach resonance two fermionic atoms with opposite spin can form a bosonic molecule. Varying the magnetic field one can detune the bosonic level compared to fermionic one. Doing this one can vary the ratio of the filling of fermions and bosons. We find a phase transition between the BEC/BCS phase and a fermionic Mott insulator.

Chapter 2

Ultracold atomic Physics

In this chapter we overview the main background of the physics considered in this thesis. Since it is one of our main goals to study the BCS-BEC crossover in an optical lattice, we elaborate on this theme, paying also attention to the historical development. Besides that, we discuss the various cooling techniques and the main characteristics of optical lattices. Finally, we derive the Hamiltonians that will be studied in the following chapters.

2.1 Short Overview

The history of superconductivity/superfluidity started in early 20th century when H. Kamerlingh Onnes discovered that if a metallic sample of mercury is cooled below 4.2K the resistance reduces to zero [46]. Later on, in 1938, two different groups independently observed superfluidity of bosonic Helium (^{4}He) [47, 48]. They observed that below 2.17K the viscosity vanishes. Much later it was observed that also fermionic Helium (^{3}He) becomes superfluid below a critical temperature of 2.7mK [49]. In 1938 F. London suggested a connection between superfluidity and Bose-Einstein condensation (BEC) [50], which was predicted by S.N. Bose [51] and A. Einstein [52, 53] in 1924-1925. For the first phenomenological understanding of superfluidity [54, 55] L. D. Landau received the Nobel prize in 1962. A microscopic understanding of this phenomenon was developed by N. N. Bogoliubov [56]. The theoretical understanding of superconductivity took considerably longer. It was not until 1956, that the key idea emerged by L. Cooper [57]. He showed that an arbitrary small interaction between two fermions (electrons) with opposite spin and momenta, in the presence of many other fermions (electrons) can lead to the formation of bound pairs thereby reducing the total energy. The presence of the other fermions, forming a Fermi sea, is crucial, because without them a critical interaction strength is required for the formation of a bound pair in three dimension. A microscopic theory for superconductivity, the BCS theory [58], was developed one year after L. Cooper's work by J. Bardeen, L. N. Cooper, and J. R. Schrieffer.

Even earlier than the formulation of the BCS theory, in 1954, M. R. Schafroth proposed that superconductivity occurs due to the existence of a charged Bose gas of two-electron bound states that condenses below the critical temperature [59]. This theory was not able to explain experiments, and the idea was not appreciated at that time. It was thought that there were fundamental differences rather than similarities between the BCS state and the BEC.

The first theoretical explanation of the crossover from the BCS state to the BEC was given by D. M. Eagles [60] in the late sixties, in the context of studying the BCS states at low carrier concentrations in superconducting semiconductors at $T = 0$. Only in 1980, A. Leggett reconciled the BCS picture of the Cooper pairs and Schafroth's BEC [61]. Leggett considered a dilute gas of fermions within a mean-field framework at $T = 0$ and showed that the BCS state was applicable in more general frame than just in the weakly interacting limit. In the weak coupling limit it describes the BCS theory of Cooper pairs, while in the strong coupling limit it describes the BEC state of strongly bounded pairs, which are bosonic in nature. Thus changing the interaction between fermions, there is a smooth crossover from BCS Cooper pairs to the BEC of composed bosons. Later on, P. Nozières and S. Schmitt-Rink used a diagrammatic method with a finite-ranged attraction and generalized Leggett's work for temperatures close to the critical temperature for superfluidity [62].

The discovery of high-T_c superconductivity in 1986 [63] showed that BCS theory fails to describe important regions of the phase diagram. It was shown that the size of Cooper pairs is much smaller than in conventional superconductors, but is larger than the size of tightly bound bosons on the BEC side. Due to this reason, the BCS-BEC crossover became a popular subject for investigation during the last two decades.

The first experimental realization of a Bose-Einstein condensate in dilute atomic gases for rubidium [64], lithium [65] and sodium [66], even more motivated scientists to study the BEC-BCS crossover. Cold atomic gases provide a perfect laboratory for comparing theoretical and experimental results with high accuracy. This can be considered as the starting point of this new area of research. The first experiments were performed on ultracold bosonic gases. In particular the important consequences of the Bose-Einstein condensation were investigated, which up to 1995 had remained an elusive and inaccessible phenomenon in experiment. In the last 15 years there has been significant experimental progress. These include accessing of hydrodynamic nature of collective oscillations [67, 68], the observation of the interference of matter waves [69], realization of: spinor condensates [70], Josephson like effects [71, 72], superfluid-Mott insulator transition [1, 73–76], Hanbury-Brown-Twiss effect [77], and many other phenomenon.

It did not take too long, after the first realization of the BEC until ultracold fermionic gases were studied experimentally as well. The first important results of quantum degeneracy in trapped Fermi gases were obtained in 1999 by the JILA group [78] and later on by other groups [4, 17]. More recent experimental works concentrated on studying the effect of spin imbalance on the BCS state, i.e. the case when an unequal number of fermions occupies two different spin states [79–82], the effect of periodic potentials on trapped Fermi gases [83–86] as well as mixtures of fermions with unequal masses, such as ^{6}Li and ^{40}K [87, 88]. Recently, for repulsively interacting fermions the Mott insulator were realized [89, 90].

Ultracold atomic gases also allow to perform experiments on Bose-Fermi mixtures [4–19]. One of the key questions that has been explored is the effect of fermions on the mobility of the bosons, in particular the effect of fermions on the superfluid-Mott transition for the bosons [10–16]. The most impressive result was that the time of flight experiments show dramatic loss of bosonic coherence. These results indicate that adding the fermions causes a stabilization of the Mott insulator phase.

A more detailed overview of the physics of cold atomic gases is given in the following review papers. [91–95].

2.2 Cooling Atoms

The ultracold atomic physics takes place at the lowest temperatures in the universe, which can only be reached by means of sophisticated cooling techniques in experiments. In Fig. 2.1 we compare these temperatures to the other typical temperatures in the universe. In this section, we shortly describe how such low temperatures can be reached.

One starts with a beam of atoms emerging from an oven at a temperature of about 600K, which corresponds to a speed of about 800m/s for sodium atoms (see Fig. 2.2).

12   2. Ultracold atomic Physics

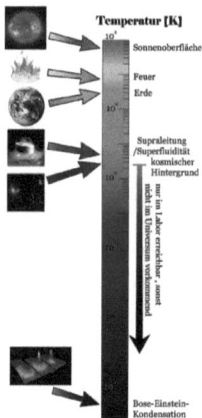

Figure 2.1: Typical temperatures in the universe (From Ref. [96]).

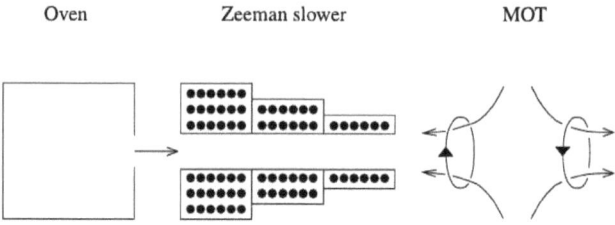

Figure 2.2: A typical experiment to cool and trap alkali atoms (From Ref. [97]).

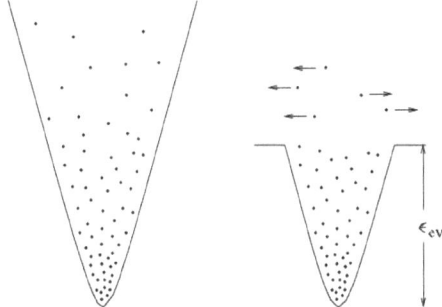

Figure 2.3: A schematic structure illustrating evaporative cooling (From Ref. [97]).

Afterwards it passes through so called Zeeman slower. In the Zeeman slower, a laser beam propagates in the direction opposite to the atomic beam, and the radiation force produced by absorption of photons retards the atoms. The velocity of atoms is reduced to about 30m/s corresponding to a temperature of order 1K. After emerging from the Zeeman slower, the atoms are already slowed down sufficiently to be captured by the magneto-optical trap (MOT), where they are further cooled by interacting with the laser light to a temperature of $100\mu K$. However, even these low temperatures are not sufficient to reach Bose-Einstein condensation, so the system has to be further cooled using evaporative cooling. This decreases the temperature to values as low as 3nK [98].

The idea of evaporative cooling is relatively simple [97, 99, 100]: by slowly reducing the height of the trap, particles with a higher energy than the edge of the trap will escape, which lowers the average energy and correspondingly the temperature in the system (see Fig. 2.3).

A more detailed discussion can be found in [97, 100, 101].

2.3 Feshbach Resonances

Interaction effects in quantum degenerate, dilute cold atomic gases can be accurately modeled by a small number of parameters characterizing the physics of two-body collisions.

The interaction potential between two neutral atoms is of the Lennard-Jones form [102]:

$$V(r) = \frac{A}{r^{12}} - \frac{B}{r^6}. \tag{2.1}$$

If one neglects the small relativistic spin-orbital interactions, the problem of describing the collision process between two atoms reduces to the solution of the Schrödinger equation for the relative motion:

$$\left(-\frac{\hbar^2}{2m_r}\nabla^2 + V(r) - \epsilon\right)\Psi(\mathbf{r}) = 0, \tag{2.2}$$

where $\mathbf{r} = \mathbf{r}_1 - \mathbf{r}_2$ is the relative coordinate of the two atoms and $m_r = m_1 m_2/(m_1 + m_2)$ is the reduced mass. For positive energies $\epsilon > 0$, the solution of Eq. (2.2) in the asymptotic region can be written as:

$$\Psi(\mathbf{r}) = e^{i\mathbf{k}\cdot\mathbf{r}} + f(k,\theta)\frac{e^{ikr}}{r}, \tag{2.3}$$

where k is the incoming relative momentum and θ is scattering angle. The scattering state at large distance appears as a superposition of the plane wave ($e^{i\mathbf{k}\cdot\mathbf{r}}$) and an outgoing spherical wave ($\frac{e^{ikr}}{r}$). $f(k,\theta)$ is by definition the scattering amplitude of the process. For $k \to 0$ the s-wave scattering amplitude tends to a constant value independent of k and θ. The quantity

$$a_s = \lim_{k\to 0} f(k,\theta) \tag{2.4}$$

is called the s-wave scattering length and plays a crucial role in the effective description of the scattering processes at low temperatures. In particular, the low-energy scattering potential between two atoms can be expressed in terms of the scattering length only

$$V(\mathbf{r} - \mathbf{r}') = \frac{4\pi\hbar^2}{m} a_s \delta(\mathbf{r} - \mathbf{r}'). \tag{2.5}$$

In the rest of this thesis we will use this representation of the interatomic potential in terms of the scattering length. Generally the use of the delta-potential leads to a divergence. In chapter 6 we will explicitly encounter this divergence, which can be dealt with by standard renormalization procedures.

The *Feshbach resonances* is an important tool to experimentally investigate cold atomic gases. It allows to tune the scattering length a_s to values much larger than the mean interatomic distance and even change its sign, by changing the external magnetic field [103]. This resonance occurs when the energy associated with an elastic scattering process

2.3 Feshbach Resonances

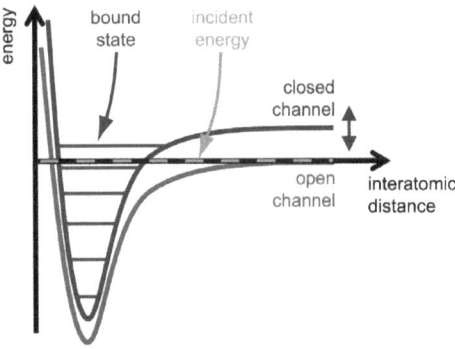

Figure 2.4: Schematic picture of a Feshbach resonance. Atoms prepared in the open channel undergo a collision at low incident energy. During the collision the open channel is coupled to the closed channel. When a bound state of the closed channel has an energy close to zero, the scattering resonance occurs. The position of the closed channel can be detuned with respect to the open one, e.g., by varying the magnetic field B. (From Ref. [92])

(*open channel*) comes close to the energy of a bound state (*closed channel*) (see Fig. 2.4). This phenomenon was first investigated in the context of reactions forming compound nuclei [104] and later on, independently, for a description of configuration interactions in multielectron atoms [105]. In the context of cold atomic physics Feshbach resonances were first used for bosonic systems [106, 107]. Due to non elastic processes the tuning of interaction strength is limited to the case of repulsive interactions [108]. In the case of fermions, due to the Pauli principle three-body losses are suppressed and this causes a greater stability of the gas [109].

Phenomenologically, a Feshbach resonance can be described by an effective pseudopotential between atoms of the open channel with the scattering length:

$$a_s(B) = a_{bg}\left(1 - \frac{\Delta B}{B - B_0}\right), \qquad (2.6)$$

where a_{bg} is the off-resonance background scattering length, B_0 is the magnetic field at resonance and ΔB is the width of the resonance. As one can see from Eq. (2.6) for the magnetic field $B_0 + \Delta B$, the scattering length $a_s(B_0 + \Delta B) = 0$. In the Fig. 2.5 we plot the scattering length as a function of the magnetic field. Now we will derive Eq. (2.6) by

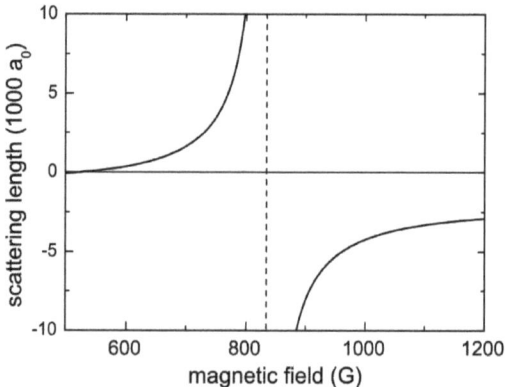

Figure 2.5: Magnetic field dependence of the scattering length between two lowest magnetic substates of ^{6}Li with the Feshbach resonance at $B_0 = 834$G and a zero crossing at $B_0 + \Delta B = 534$G. The background scatterring length is $a_{bg} = -1405 a_0$, where a_0 is the Bohr radius. (from Ref. [92])

following Ref. [97].

The space of states describing two scattering atoms can be divided into two subspaces: P, which contains the open channel and Q, which contains the closed ones. We write the state vector $|\Psi\rangle$ as the sum of its projections onto the two subspaces:

$$|\Psi\rangle = |\Psi_P\rangle + |\Psi_Q\rangle, \tag{2.7}$$

where $|\Psi_P\rangle = \hat{\mathcal{P}}|\Psi\rangle$ and $|\Psi_Q\rangle = \hat{\mathcal{Q}}|\Psi\rangle$. Here $\hat{\mathcal{P}}$ and $\hat{\mathcal{Q}}$ are projection operators, that fulfill the following relations: $\hat{\mathcal{P}} + \hat{\mathcal{Q}} = \mathbb{1}$ and $\hat{\mathcal{P}}\hat{\mathcal{Q}} = \hat{\mathcal{Q}}\hat{\mathcal{P}} = 0$. From this it follows that

$$|\Psi\rangle = (\hat{\mathcal{P}} + \hat{\mathcal{Q}})^2 |\Psi\rangle = \left(\hat{\mathcal{P}}^2 + \hat{\mathcal{Q}}^2 \right) |\Psi\rangle = \hat{\mathcal{P}} |\Psi_P\rangle + \hat{\mathcal{Q}} |\Psi_Q\rangle. \tag{2.8}$$

Now let us multiply the Schrödinger equation $\hat{\mathcal{H}}|\Psi\rangle = E|\Psi\rangle$ on the left by $\hat{\mathcal{P}}$ and $\hat{\mathcal{Q}}$, and use Eq. (2.8). We thus obtain:

$$(E - \hat{\mathcal{H}}_{PP})|\Psi_P\rangle = \hat{\mathcal{H}}_{PQ}|\Psi_Q\rangle, \tag{2.9}$$
$$(E - \hat{\mathcal{H}}_{QQ})|\Psi_Q\rangle = \hat{\mathcal{H}}_{QP}|\Psi_P\rangle. \tag{2.10}$$

2.3 Feshbach Resonances

Here we define

$$\hat{\mathcal{H}}_{PP} = \hat{P}\hat{\mathcal{H}}\hat{P}, \quad \hat{\mathcal{H}}_{QQ} = \hat{Q}\hat{\mathcal{H}}\hat{Q}, \quad \hat{\mathcal{H}}_{PQ} = \hat{P}\hat{\mathcal{H}}\hat{Q} \quad \text{and} \quad \hat{\mathcal{H}}_{QP} = \hat{Q}\hat{\mathcal{H}}\hat{P}. \tag{2.11}$$

Combining Eqs. (2.9) and (2.10) we obtain that

$$\left(E - \hat{\mathcal{H}}_{PP} - \hat{\mathcal{H}}'_{PP}\right)|\Psi_P\rangle = 0, \tag{2.12}$$

where

$$\hat{\mathcal{H}}'_{PP} = \hat{\mathcal{H}}_{PQ}\left(E - \hat{\mathcal{H}}_{QQ} + i\delta\right)^{-1}\hat{\mathcal{H}}_{QP}. \tag{2.13}$$

Adding a positive infinitesimal imaginary part δ in the denominator ensures that the scattered wave contains only outgoing terms.

It is convenient to divide $\hat{\mathcal{H}}_{PP} + \hat{\mathcal{H}}_{QQ}$ into a term $\hat{\mathcal{H}}_0$ independent of the separation of the two atoms, and an interaction contribution. We write

$$\hat{\mathcal{H}}_{PP} = \hat{\mathcal{H}}_0 + \hat{U}_1, \tag{2.14}$$

where $\hat{\mathcal{H}}_0$ is the sum of the kinetic energy of the relative motion and the hyperfine and Zeeman terms. $\hat{U}_1$ is the interaction term for the P subspace. The total atom-atom interaction in the subspace of the open channel is given by

$$\hat{U} = \hat{U}_1 + \hat{U}_2, \tag{2.15}$$

where

$$\hat{U}_2 = \hat{\mathcal{H}}'_{PP}. \tag{2.16}$$

The quantity which characterizes the elastic scattering is the T-matrix $T(\mathbf{k}', \mathbf{k}, \hbar^2 k^2/2m_r)$. Here $\mathbf{k}$ and $\mathbf{k}'$ are incoming and outgoing relative momentum and $m_r = m/2$ is the reduced mass. There is the following relation, between the T-matrix and scattering length a_s:

$$T(0,0,0) = \frac{4\pi\hbar^2}{m}a_s. \tag{2.17}$$

One can calculate the T-matrix, by solving the Lippmann-Schwinger operator equation:

$$\hat{T} = \hat{U} + \hat{U}\hat{G}_0\hat{T}, \tag{2.18}$$

where

$$\hat{G}_0 = (E - \hat{\mathcal{H}}_0 + i\delta)^{-1} \tag{2.19}$$

is the Green's function for the Schrödinger equation. The formal solution of Eq. (2.18) is

$$\hat{T} = \left(1 - \hat{U}\hat{G}_0\right)^{-1}\hat{U} = \hat{U}\left(1 - \hat{G}_0\hat{U}\right)^{-1}. \tag{2.20}$$

One can show that

$$\hat{T} = \hat{T}_P + \left(1 - \hat{U}_1\hat{G}_0\right)^{-1}\hat{U}_2\left(1 - \hat{G}_0\hat{U}\right)^{-1}. \tag{2.21}$$

Here

$$\hat{T}_P = \left(1 - \hat{U}_1\hat{G}_0\right)^{-1}\hat{U}_1 \tag{2.22}$$

is the T-matrix in P subspace if transitions into the Q subspace are neglected.

We now apply the above results, and we begin by considering the contribution that are first order in $\hat{U}_2$. This is equivalent to replacing $\hat{U}$ by $\hat{U}_1$ in Eq. (2.21), which gives

$$\hat{T} = \hat{T}_P + \left(1 - \hat{U}_1\hat{G}_0\right)^{-1}\hat{U}_2\left(1 - \hat{G}_0\hat{U}_1\right)^{-1}. \tag{2.23}$$

The matrix elements between plane-wave states are given by

$$\langle \mathbf{k}'|\hat{T}|\mathbf{k}\rangle = \langle \mathbf{k}'|\hat{T}_P|\mathbf{k}\rangle + \langle \mathbf{k}'; U_1, -|\hat{U}_2|\mathbf{k}; U_1, +\rangle, \tag{2.24}$$

where $|\mathbf{k}; U_1, +\rangle = \left(1 - \hat{G}_0\hat{U}_1\right)^{-1}|\mathbf{k}\rangle$ and $\langle \mathbf{k}'; U_1, -| = \langle \mathbf{k}'|\left(1 - \hat{U}_1\hat{G}_0\right)^{-1}$.

Now we neglect the coupling between the open channels. In this case we can neglect the difference between the scattering states with incoming and outgoing spherical waves and define this state as $|\psi_0\rangle$. Let us define the states of the closed channel as $|\psi_n\rangle$ ($n \neq 0$). Then we obtain from Eqs. (2.13), (2.16), (2.17) and (2.24) that the scattering length is

2.3 Feshbach Resonances

given by the following expression:

$$\frac{4\pi\hbar^2}{m}a_s = \frac{4\pi\hbar^2}{m}a_P + \sum_{n\neq 0}\frac{|\langle\psi_n|\hat{\mathcal{H}}_{QP}|\psi_0\rangle|^2}{E_{th} - E_n}, \quad (2.25)$$

where E_{th} is the *threshold energy* and a_P is the scattering length if the coupling between the open and the closed channel is neglected and the sum n is over all states of the closed channel.

If E_{th} is close to one of the states of the closed channel E_m, then the contribution of all other terms in the sum in Eq. (2.25), except of the resonance term $n = m$, will vary slowly with energy and one can neglect this dependence. So we will obtain:

$$\frac{4\pi\hbar^2}{m}a_s = \frac{4\pi\hbar^2}{m}a_{bg} + \frac{|\langle\psi_n|\hat{\mathcal{H}}_{QP}|\psi_0\rangle|^2}{E_{th} - E_{res}}. \quad (2.26)$$

This expression shows how the scattering length depends on the energy. We assume that for $B = B_0$, the denominator vanishes, i.e. $E_{th} = E_{res}$. This enables us to expand $E_{th} - E_{res}$ around this value of the magnetic field

$$E_{th} - E_{res} \simeq C(B - B_0). \quad (2.27)$$

Putting Eq. (2.27) in Eq. (2.26) we directly obtain Eq. (2.6).

In the end of this section it should be mentioned that there exist two different types of the Feshbach resonances: *broad* and *narrow*. For the broad resonance, single channel calculations is in excellent agreement with the outcomes of coupled-channel calculations over a wide range of the magnetic field. For the narrow resonance, such a mapping can be realized only for a very narrow region of the magnetic field, which experimentally is not accessible [110, 111].

In a Fermi gas the distinction between broad and narrow resonances involves the comparison of the wave vector to the length scale associated with the inverse width $(R^\star)^{-1}$ of the resonance. If the condition

$$k_F|R^\star| \ll 1 \quad (2.28)$$

is satisfied, the effective range is irrelevant at the many body level and the properties of the gas can be described by the dimensionless parameter $k_F a_s$ only. This corresponds to a broad Feshbach resonance. On the contrary, for a narrow Feshbach resonance, $k_F|R^\star| \gtrsim 1$, the effective range becomes a relevant scale of the problem [112–114].

A broad Feshbach resonance for ^{40}K atoms take place at $B \simeq 202.1$G [115, 116] and for ^{6}Li atoms at $B = 834$G [117–120]. In both of these cases $k_F|R^*| \lesssim 0.01$. A Narrow Feshbach resonance occurs for ^{6}Li atoms at $B = 543.23$G [121] and in this case $k_F|R^*| \gtrsim 1$.

To conclude, the Feshbach resonance is a good tool to change the interaction strength between ultracold atoms over a wide range. It even allows to change the sign of the interaction. More details about Feshbach resonances can be found in the following references: [95, 97, 111, 122–125]

2.4 BCS-BEC Crossover

As we already mentioned in section 2.1, the BCS-BEC crossover is one of the challenging problems in condensed matter physics. In this section we will briefly overview this huge topic.

If the scattering length is tuned to small positive values, fermions with different spin will form a dimer and further increase of the interaction strength will lead to the formation of a bosonic gas of molecules. Such molecules can be obtained by cooling a gas at a positive value of the scattering length. Alternative way to obtain such a molecule is, first cool gas, obtain BCS state and then tune interaction. During this process BCS-BEC crossover can be observed. Bose-Einstein condensation of pairs of atoms at very low temperatures was observed by several groups [119, 120, 126–128]. Later on the properties of these systems also were studied in the BCS regime [117, 128, 129].

Up to now no exact analytical solution of the many body problem along the whole BCS-BEC crossover exists, but there is a general agreement at a qualitative level. The BCS, weak attractive limit, occurs for $1/k_f a_s \ll -1$, while the BEC occurs for $1/k_f a_s \gg 1$, and the crossover occurs in the region for $-1 < 1/k_f a_s < 1$.

In Fig. 2.6 we plot the qualitative phase diagram for the BEC-BCS crossover. If temperature is relatively high, the normal state with weak attraction is a Fermi liquid. If one increases the interaction, fermion pair-formation will take place. These two regions are separated by the pair formation temperature T_{pair} (the pink (dark) line in the Fig. 2.6). If one continues increasing the interaction then the system will smoothly evolve to a molecular Bose liquid.

2.4 BCS-BEC Crossover

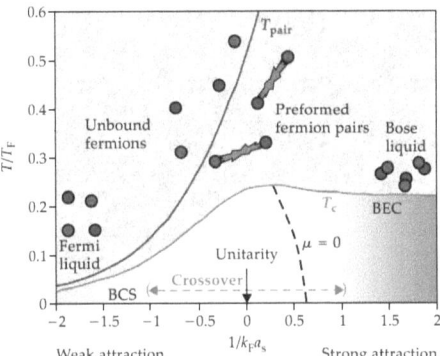

Figure 2.6: The qualitative phase diagram for the BEC-BCS crossover. The pink (dark) line corresponds to the pair formation temperature T_{pair}, the blue (light) line corresponds to the critical temperature T_c. Bellow this temperature phase coherence between pairs is established. The unitarity limit occurs when $1/k_f a_s = 0$, where k_f is the Fermi momentum and a_s is the scattering length. (From Ref. [130])

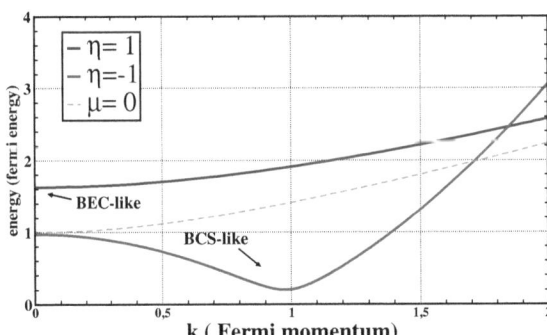

Figure 2.7: Energy spectrum E_k from the Leggett theory for: (i) $\eta = 1/k_f a_s = 1$, $\Delta/E_F = 1.36$ and $\mu/E_f = -0.8$ (blue line); (ii) $\eta = -1$, $\Delta/E_F = 0.2$ and $\mu/E_f = 0.95$ (red line), (iii) $\eta = 0.55$, $\Delta/E_F = 1.05$ and $\mu/E_f = 0$ (green dashed line). On the axes energy and momentum are measured in units of the Fermi energy and the Fermi momentum, respectively. (From Ref. [131])

To observe the BEC-BCS crossover one has to cool the system to much lower temperatures $T < T_c$. The difference between the critical temperature T_c and the pair-formation temperature T_{pair} is quite small on the BCS side, and is large on the BEC side. It means that in the BCS limit, pair formation and developing phase coherence take place at the similar temperature, while on the BEC side, first formation of the fermion pairs (diatomic molecules) takes place at a temperature T_{pair} and only when the system is cooled below the critical temperature $T < T_c$ phase coherence will emerge.

There are two more important points in this phase diagram, which are worth mentioning. One of them is the *unitary* limit, at which the scattering length diverges, i.e. $1/k_f a_s = 0$. In this case, as it was first noticed by T. L. Ho [132], there is no energy or length scale related to the interaction. Thus this is a special point because the behavior of the system does not depend on the interparticle potential. The only scale which is relevant is the Fermi energy E_F. Another important limit is the vanishing of the chemical potential ($\mu = 0$), which takes place beyond the unitarity limit at zero temperature. On the BCS side $\mu > 0$ and the energy gap in the elementary excitation spectrum is equal to the superconducting order parameter and occurs for the finite momentum (see Fig. 2.7), while on the BEC side, $\mu < 0$ and the energy gap in elementary excitation spectrum does not only depend on the superconducting order parameter, but also depends on the chemical potential and is equal to $\sqrt{\Delta^2 + \mu^2}$, where Δ and μ are the superconducting order parameter and the chemical potential, respectively. The gap occurs at zero momentum in contrast to the BCS side (see Fig. 2.7). Therefor, the point where chemical potential vanishes is the de facto separation between the BCS and the BEC.

The previous discussion was for the case of a balanced mixture of fermions, with equal masses and with s-wave scattering. But the BCS-BEC crossover physics is much richer. One can consider a mixture of an unequal number of fermions [79–82] or a mixture of fermions with unequal masses [87, 88]. Also, one can study the effect of disorder in the BCS-BEC evolution or p-wave scattering. Ultracold atomic gases also enable us to study mixtures of three hyperfine states and study BCS-BEC crossover in such a systems, in particular *color superconductivity* and the formation of *trions* [133–136].

2.5 Optical Lattices

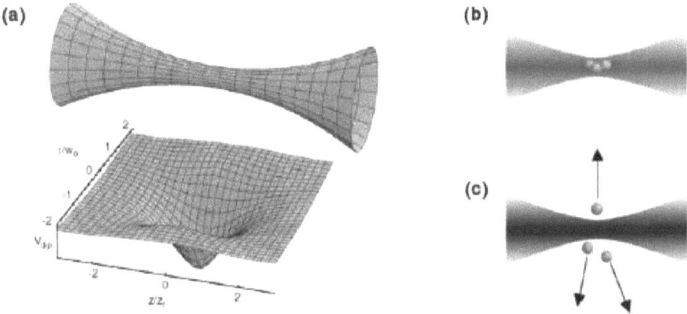

Figure 2.8: (a) A Gaussian laser beam together with the corresponding trapping potential for a red detuned laser. (b) A red detuned laser beam leads to an attractive dipole potential, whereas a blue detuned laser beam leads to repulsive potential (c). (From Ref. [93])

2.5 Optical Lattices

In this section we consider cold atomic gases in optical lattices, which are created by standing light waves. Such a systems is a good analog to conventional condensed matter systems. One can trap millions of atoms in a periodic optical lattice, whose lattice geometry and lattice depth are under full control of the experimentalist. As we have already discussed in section 2.3, one can also control the interaction strength. That means that we can fully control parameters of the effective tight binding model (in detail we consider this in section 2.6).

There exist two fundamental forces on the atoms due to the coherent light field [101, 137]. One of them is the *Doppler force*, which is dissipative in nature and relies on radiation pressure and spontaneous emission. This can be used to efficiently laser cool a gas of atoms. The second one is the *dipole force*, which creates a purely conservative potential, in which the atoms can move. If the atoms are sufficiently cold this force can be used to trap them [138, 139].

The interaction between the external electrical field with the atoms in the dipole approximation is given by

$$\mathcal{H}' = -\vec{d} \cdot \vec{\mathcal{E}}, \tag{2.29}$$

where $\vec{d}$ is the dipole moment of the atom and $\vec{\mathcal{E}}$ is the external electric field. The resulting

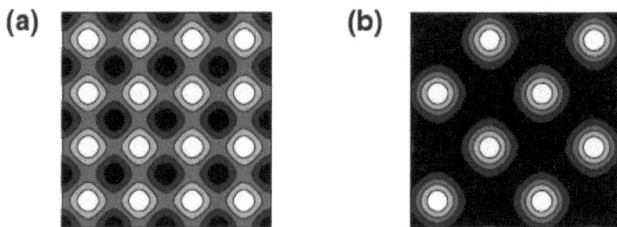

Figure 2.9: Two-dimensional optical lattice potentials for orthogonal polarization (a) and parallel polarization with time phase $\phi = 0$ (From Ref. [93]).

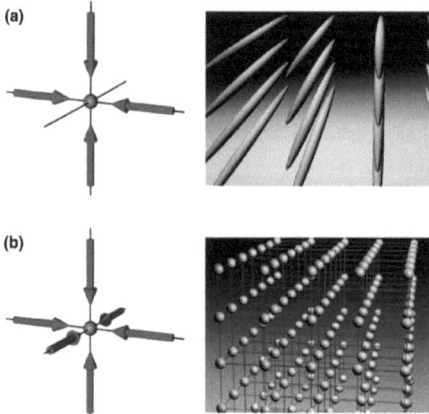

Figure 2.10: Two-dimensional (a) and three-dimensional (b) optical lattice potentials. For a two-dimensional optical lattice, the atoms are confined to an array of tightly confining one-dimensional potential tubes, while in the three-dimensional case the optical lattice can be approximated by a three-dimensional simple cubic array of tightly confining harmonic oscillator potentials at each site (From Ref. [93]).

2.5 Optical Lattices

dipole potential is given by (see e.g. [97]):

$$V_{\text{dip}} = -\frac{1}{2}\langle \vec{d} \cdot \vec{\mathcal{E}} \rangle_t = \frac{\hbar \delta \Omega_R^2}{\delta^2 + \Gamma_e^2/4}, \qquad (2.30)$$

where $\langle \ldots \rangle_t$ denotes time average, $1/\Gamma_e$ is the lifetime of the exited state, the *Rabi frequency* is defined as

$$\Omega_R = |\langle e|\vec{d} \cdot \vec{\mathcal{E}}|g\rangle|/\hbar \qquad (2.31)$$

and the *detuning* given by the equation:

$$\delta = \omega_L - (E_e - E_g)/\hbar. \qquad (2.32)$$

$|g\rangle$ and $|e\rangle$ are the ground state and the excited state and E_g and E_e are the corresponding energies, respectively.

From the Eq. (2.30) it is clear that a *red detuned* ($\delta < 0$) laser beam leads to an attractive dipole potential, whereas a *blue detuned* ($\delta > 0$) laser beam leads to a repulsive potential (see Fig. 2.8). This means that a red detuning laser can be used for trapping cold atoms.

Around the intensity maximum a potential depth minimum occurs for a red detuned laser beam. Close to this point, the dipole potential can be approximated by a harmonic potential of the form:

$$V_{\text{dip}} \simeq V_0 \left[1 - 2\left(\frac{r}{w_0}\right)^2 - \left(\frac{z}{z_R}\right)^2\right], \qquad (2.33)$$

where w_0 is the *beam waist*, which typically is 100μm and $z_R = \pi r_0^2/\lambda$ is called the *Rayleigh range*. r is the radial distance from the center axis of the beam and z is the axial distance from the beam's narrowest point.

To form a periodic potential one can use two overlapping counterpropagating laser beams with wavelength λ. The standing wave with period $\lambda/2$ will be formed due to the interference between the two laser beams. Depending on the number of the lasers, one can obtain one, two, or three dimensional periodic lattices [140]. If one interferes two laser beams less than 180°, then one can realize a periodic potential with larger period [141, 142]). In the last part of this section we will consider the simplest one-, two- and three-dimensional lattices.

One can obtain the simplest possible lattice by overlapping two counterpropagating focused Gaussian beams. In this case the electrical field along the x axis is given by:

$$\mathcal{E}_x(r,z,t) = \mathcal{E}_0(r,z)\left(\sin(qz - \omega t) + \sin(qz + \omega t)\right) = 2\mathcal{E}_0(r,z)\sin(qz)\sin(\omega t), \quad (2.34)$$

where $q = 2\pi/\lambda$ is wave vector of the laser light. From here one can directly obtain that:

$$V(r,z) \simeq 4V_0 \left[1 - 2\left(\frac{r}{w_0}\right)^2 - \left(\frac{z}{z_R}\right)^2\right] \sin^2(qz). \quad (2.35)$$

As one can clearly see from this expression one obtains one-dimensional periodic lattice with period $\lambda/2$.

A periodic potential in two-dimensions can be formed in the same way, but in this case instead of two laser beams one have to use four laser beams. Here we will consider the case when two standing waves are perpendicular to each other, i.e. a rectangular optical lattice. Neglecting the Gaussian beam profile we obtain :

$$V(y,z) \simeq V_{\text{lat}} \left[\cos^2(qy) + \cos^2(qz) + 2\vec{e}_1 \cdot \vec{e}_2 \cos(\phi) \cos(qy) \cos(qz)\right], \quad (2.36)$$

where $\vec{e}_1$ and $\vec{e}_2$ denote the polarization vectors of the perpendicular laser field and ϕ is the temporal phase between them. From Eq. (2.36), it is clear that if the polarization vectors are not perpendicular to one another, then changing the temporal phase one can modify the lattice structure (See Fig. 2.9).

In such a periodic potential, the atoms are confined to arrays of tightly confined one-dimensional tubes (see Fig. 2.10a). Usually the harmonic trapping frequencies along the tubes are very weak in the experiment (order of $10 - 200$ Hz), while in the radial direction they are much larger (order of 100 kHz) [93].

Using six laser beams, a three-dimensional lattice can be formed (see Fig. 2.10a). If the polarizations of the different standing waves are perpendicular to each other, then all standing waves are independent from each other and the optical potential can be described by:

$$V(x,y,z) \simeq -V_x^2 e^{-2(y^2+z^2)/w_x^2}\sin^2(q_x x) - V_y^2 e^{-2(x^2+z^2)/w_y^2}\sin^2(q_y y) - V_z^2 e^{-2(y^2+z^2)/w_y^2}\sin^2(q_z z). \quad (2.37)$$

Here V_α and q_α are the potential depth and the wave vectors for the standing waves in α direction, respectively. At the center of the trap, for distances much smaller than the beam waist, the trap potential can be approximated as the sum of the periodic lattice potential and the external harmonic confinement due to the Gaussian laser beam profile:

$$V(x,y,z) \simeq -V_x^2 \sin^2(q_x x) - V_y^2 \sin^2(q_y y) - V_z^2 \sin^2(q_z z) + \frac{m}{2}\left(\omega_x^2 x^2 + \omega_y^2 y^2 + \omega_z^2 z^2\right), \quad (2.38)$$

where the effective trapping frequencies of the harmonic confinement are

$$\omega_x^2 = \frac{4}{m}\left(\frac{V_y}{w_y^2} + \frac{V_z}{w_z^2}\right); \qquad \omega_{y,z}^2 = \text{cycl.perm.} \quad (2.39)$$

More details about optical lattices can be found in Ref. [93] and references therein.

2.6 The Hubbard Model

2.6.1 The Fermionic Hubbard Model

The Hamiltonian that describes fermions in the periodic optical lattice in second quantized form is given by

$$\hat{\mathcal{H}} = \sum_\sigma \int d^3\mathbf{r}\, \hat{\Psi}_{f\sigma}^\dagger(\mathbf{r}) \left(-\frac{\hbar^2 \nabla^2}{2m_f} + V_f(\mathbf{r})\right) \hat{\Psi}_{f\sigma}(\mathbf{r}) + \int \hat{\Psi}_{f\downarrow}^\dagger(\mathbf{r}) \hat{\Psi}_{f\uparrow}^\dagger(\mathbf{r}) \frac{4\pi\hbar^2 a_f}{m_f} \hat{\Psi}_{f\uparrow}(\mathbf{r}) \hat{\Psi}_{f\downarrow}(\mathbf{r}), \quad (2.40)$$

where $\hat{\Psi}_{f\sigma}^\dagger(\mathbf{r})$ is the creation operator for a fermion with spin σ at point $\mathbf{r}$. m_f is its mass, a_f is the s-wave scattering length and $V_f(\mathbf{r})$ denotes the periodic potential.

In the presence of the periodic potential $V_f(\mathbf{r}+\mathbf{R}) = V_f(\mathbf{r})$, it is convenient to express the fermionic creation operators $\hat{\Psi}_{f\sigma}^\dagger(\mathbf{r})$ using Wannier functions:

$$\hat{\Psi}_{f\sigma}^\dagger = \sum_{i,l} \hat{c}_{i\sigma l}^\dagger w_{l,x}^f(x-x_i) w_{l,y}^f(y-y_i) w_{l,z}^f(z-z_i), \quad (2.41)$$

where $\hat{c}_{i\sigma l}^\dagger$ is fermionic creation operators at site $\mathbf{r}_i = (x_i, y_i, z_i)$ and $w_l^f(\mathbf{r}-\mathbf{r}_i) = w_{l,x}^f(x-x_i) w_{l,y}^f(y-y_i) w_{l,z}^f(z-z_i)$ is the Wannier function for a localized particle in the lth energy band.

For temperatures and interactions small compared to the band gap, only the lowest band will be occupied. That means that the sum over the band indices l is reduced to $l = 0$ and we can drop the band index.

Inserting Eq. (2.41) into equation (2.40) one obtains:

$$\hat{\mathcal{H}} = -t_f \sum_{\langle ij \rangle, \sigma} \hat{c}^\dagger_{i\sigma} \hat{c}_{j\sigma} + U_f \sum_i \hat{n}^f_{i\uparrow} \hat{n}^f_{i\downarrow}, \qquad (2.42)$$

where

$$t_f = \int d^3\mathbf{r} w^f_x(x - x_i + a) w^f_y(y - y_i) w^f_z(z - z_i) \frac{\hbar^2 \nabla^2}{2m_f} w^f_x(x - x_i) w^f_y(y - y_i) w^f_z(z - z_i) \qquad (2.43)$$

is the hopping amplitude between nearest neighbor sites $\langle ij \rangle$ and

$$U_f = \frac{4\pi \hbar^2 a_f}{m_f} \int dx |w^f_x(x - x_i)|^4 \int dy |w^f_y(y - y_i)|^4 \int dz |w^f_z(z - z_i)|^4 \qquad (2.44)$$

is the on-site Hubbard interaction.

For a deep optical lattice, one can approximate the Wannier functions by Gaussian ones, i.e:

$$w^f_\alpha(\alpha - \alpha_i) = \frac{e^{-(\alpha - \alpha_i)^2 / 2l_f^2}}{\pi^{1/4} l_f^{1/2}}. \qquad (2.45)$$

Here $\alpha = x, y, z$, and

$$l_f = \frac{a}{\pi (V_0^f / E_r^f)^{1/4}}, \qquad (2.46)$$

where $E_r^f = h^2 / 2\lambda^2 m_f$ is a recoil energy, V_0^f is the laser potential strength and λ is the wavelength of laser.

According to this approximation we obtain that (details of the derivation one can see in Appendix A):

$$U_f \simeq \sqrt{\frac{8}{\pi}} k a_f E_r^f \left(\frac{V_0^f}{E_r^f} \right)^{3/4}, \qquad (2.47)$$

while due to the much weaker side wings of the Gaussian function, the hopping amplitude t_f may be underestimated by almost order of the magnitude in this way. The correct value for the hopping amplitude t_f can be obtained from the width $W \to 4t_f$ of the lowest band in the 1D Mathieu equation [143]:

$$t_f \simeq \frac{4}{\sqrt{\pi}} E_r^f \left(\frac{V_0^f}{E_r^f} \right)^{3/4} \exp\left[-2\sqrt{\frac{V_0^f}{E_r^f}} \right]. \qquad (2.48)$$

2.6.2 The Bose-Fermi Hubbard Model

Now we will consider a mixture of bosons and two-component fermions in a sufficiently deep optical lattice. The Hamiltonian in second quantized form is given by:

$$\hat{\mathcal{H}} = \hat{T}_f + \hat{T}_b + \hat{V}_f + \hat{V}_b + \hat{W}_{ff} + \hat{W}_{bb} + \hat{W}_{fb}, \tag{2.49}$$

with the individual terms

$$\hat{T}_f = -\sum_\sigma \int d^3\mathbf{r}\, \hat{\Psi}_{f\sigma}^\dagger(\mathbf{r}) \frac{\hbar^2 \nabla^2}{2m_f} \hat{\Psi}_{f\sigma}(\mathbf{r}), \tag{2.50}$$

$$\hat{T}_b = -\int d^3\mathbf{r}\, \hat{\Psi}_b^\dagger(\mathbf{r}) \frac{\hbar^2 \nabla^2}{2m_b} \hat{\Psi}_b(\mathbf{r}), \tag{2.51}$$

$$\hat{W}_{ff} = \int \hat{\Psi}_{f\downarrow}^\dagger(\mathbf{r}) \hat{\Psi}_{f\uparrow}^\dagger(\mathbf{r}) \frac{4\pi\hbar^2 a_f}{m_f} \hat{\Psi}_{f\uparrow}(\mathbf{r}) \hat{\Psi}_{f\downarrow}(\mathbf{r}), \tag{2.52}$$

$$\hat{W}_{bb} = \frac{1}{2} \int \hat{\Psi}_b^\dagger(\mathbf{r}) \hat{\Psi}_b^\dagger(\mathbf{r}) \frac{4\pi\hbar^2 a_b}{m_b} \hat{\Psi}_b(\mathbf{r}) \hat{\Psi}_b(\mathbf{r}), \tag{2.53}$$

$$\hat{W}_{fb} = \sum_\sigma \int \hat{\Psi}_{f\sigma}^\dagger(\mathbf{r}) \hat{\Psi}_b^\dagger(\mathbf{r}) \frac{2\pi\hbar^2 a_{fb}}{m_r} \hat{\Psi}_b(\mathbf{r}) \hat{\Psi}_{f\sigma}(\mathbf{r}), \tag{2.54}$$

$$\hat{V}_f = \sum_\sigma \int d^3\mathbf{r}\, \hat{\Psi}_{f\sigma}^\dagger(\mathbf{r}) V_f(\mathbf{r}) \hat{\Psi}_{f\sigma}(\mathbf{r}), \tag{2.55}$$

$$\hat{V}_b = \int d^3\mathbf{r}\, \hat{\Psi}_b^\dagger(\mathbf{r}) V_b(\mathbf{r}) \hat{\Psi}_b(\mathbf{r}). \tag{2.56}$$

Here $\hat{\Psi}_b^\dagger(\mathbf{r})$ is the creation operator of a boson at point $\mathbf{r}$. a_b and a_{fb} are the s-wave scattering lengths for Bose-Bose and Bose-Fermi interactions, respectively. m_b is the mass of the bosons and $m_r = m_f m_b/(m_f + m_b)$. $V_b(\mathbf{r})$ denotes the periodic potential for bosons.

As we discussed above, in the presence of a strong optical lattice, the field operators can be expanded in terms of the single-particle Wannier functions localized at (x_i, y_i, z_i):

$$\hat{\Psi}_b^\dagger = \sum_{i,l} \hat{b}_{i,l}^\dagger w_{l,x}^b(x - x_i) w_{l,y}^b(y - y_i) w_{l,z}^b(z - z_i), \tag{2.57}$$

where $\hat{b}_{i,l}$ are bosonic creation operators at site (x_i, y_i, z_i).

Inserting Eq. (2.41) and (2.57) into equations (2.50-2.56) one obtains:

$$\hat{\mathcal{H}} = -t_f \sum_{\langle ij\rangle,\sigma} \hat{c}_{i\sigma}^\dagger \hat{c}_{j\sigma} - t_b \sum_{\langle ij\rangle} \hat{b}_i^\dagger \hat{b}_j + U_f \sum_i \hat{n}_{i\uparrow}^f \hat{n}_{i\downarrow}^f + \frac{U_b}{2} \sum_i \hat{n}_i^b(\hat{n}_i^b - 1) + U_{fb} \sum_i \hat{n}_i^f \hat{n}_i^b - \sum_i \mu_i^f \hat{n}_i^f - \sum_i \mu_i^b \hat{n}_i^b. \tag{2.58}$$

One can easily show that (see detailed derivation in Appendix A):

$$t_b \simeq \frac{4}{\sqrt{\pi}} E_r^b \left(\frac{V_0^b}{E_r^b}\right)^{3/4} \exp\left[-2\sqrt{\frac{V_0^b}{E_r^b}}\right], \quad (2.59)$$

$$U_b \simeq \sqrt{\frac{8}{\pi}} k a_b E_r^b \left(\frac{V_0^b}{E_r^b}\right)^{3/4}, \quad (2.60)$$

$$U_{fb} \simeq \frac{4}{\sqrt{\pi}} k a_{fb} E_r^b \frac{1 + m_b/m_f}{(1 + \sqrt{m_b/m_f})^{3/2}} \left(\frac{V_0^b}{E_r^b}\right)^{3/4}, \quad (2.61)$$

where $E_r^b = \hbar^2/2\lambda^2 m_b$ is a recoil energy, V_0^b is the laser potential strength and λ is the wavelength of laser.

At sufficiently low temperatures, the only relevant scattering processes occur within the s-wave subspace. Consequently the interaction strength between two spinless fermions vanishes and the Hamiltonian for a mixture of bosons and spinless fermions takes on the form

$$\hat{\mathcal{H}} = -t_f \sum_{\langle ij \rangle} \hat{c}_i^\dagger \hat{c}_j - t_b \sum_{\langle ij \rangle} \hat{b}_i^\dagger \hat{b}_j + \frac{U_b}{2} \sum_i \hat{n}_i^b(\hat{n}_i^b - 1) + U_{fb} \sum_i \hat{n}_i^f \hat{n}_i^b - \sum_i \mu_i^f \hat{n}_i^f - \sum_i \mu_i^b \hat{n}_i^b. \quad (2.62)$$

So far in this section a periodic potential was considered, but as we have already mentioned in section 2.5, spatial inhomogeneity due to the harmonic confinement potential is always present, leading to a spatially varying local density. However, if harmonic confinement is shallow than we can neglect effect of the trap and during calculating model parameters consider it periodic. It is important to realize that shallow trap means that laser potential strength $V_0^{f(b)}$ is much smaller than the band gap, which is usually large. This condition is fulfilled for all calculations considered in this thesis.

In case of a tight trap one cannot any more consider the system as a periodic one and correspondingly Bloch's theorem breaks down. Correspondingly, the way how we calculate the model parameters in Eqs. (2.47), (2.48), (2.59 - 2.61) is not valid any more and in this case one has to consider that the lattice sites in the harmonic trap are not equivalent to each other.

Chapter 3

Method

3.1 Dynamical Mean-Field Theory (DMFT)

As we mentioned in the introduction, ultracold atomic gases of fermions can be well described by the Hubbard model (2.42). In spite of the simplicity of this model it can be solved exactly only in one spatial dimension, where the Bethe Ansatz can be applied. On the other hand in higher dimensions, only approximate analytical solutions are available, e.g. mean-field or perturbative calculations, whose reliability is typically limited to the weak ($U_f/t_f \ll 1$) or strong ($U_f/t_f > 0$) coupling regime. Due to this numerical calculations are unavoidable to overcome these limitations.

One of the ways to study such a system is using quantum Monte Carlo (QMC) simulations [144–146]. The disadvantage of this method is that one can study only small size systems and then extrapolate the obtained results to the thermodynamic limit. Also for fermions it has minus sign problem. One of the methods which reliably describes thermodynamical systems on a lattice from weak to strong coupling is the *Dynamical Mean-Field Theory* (DMFT) [147–155]. The only small parameter in this method is $1/z$, where z is the lattice coordination number. In this section we will consider the DMFT in detail.

The main idea of the DMFT approach is to map the quantum lattice problem with many degrees of freedom onto a single site - the *"impurity site"* - coupled self-consistently to a non-interacting bath. To derive the self-consistency equations for this model, we use the "cavity method" [148]: one considers a single site of the lattice and integrates out the remaining degrees of freedom on all other sites (schematically the DMFT is depicted in Fig. 3.1). To derive the self-consistency relations, we use the path integral formalism.

In this section we will consider the DMFT for fermions. Ultracold fermions in an optical

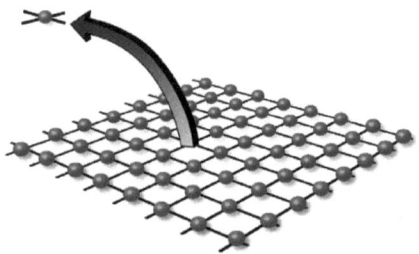

Figure 3.1: Schematic picture of Dynamical Mean-Field Theory (DMFT). Within the DMFT approach the full many-body lattice problem is replaced by a single-site problem, which is self-consistently coupled to a fermionic bath.

lattice are well described by the Hubbard model:

$$\hat{\mathcal{H}} = -t_f \sum_{\langle i,j\rangle,\sigma} \hat{c}^\dagger_{i\sigma}\hat{c}_{j\sigma} + U_f \sum_i \hat{n}^f_{i\uparrow}\hat{n}^f_{i\downarrow} - \sum_{i,\sigma} \mu_{\sigma f}\hat{n}^f_{i\sigma}, \qquad (3.1)$$

where $\hat{c}^\dagger_{i\sigma}$ is the fermionic creation operator at site i with spin σ, while $\hat{n}^f_{i\sigma} = \hat{c}^\dagger_{i\sigma}\hat{c}_{i\sigma}$ denotes the number operator for spin σ fermions at site i. $\mu_{\sigma f}$ is the chemical potential for fermions with spin σ and U_f is a local interaction between fermions with different spins. $\langle i,j\rangle$ denotes summation over nearest neighbors and t_f is the tunneling amplitude for fermions.

The important point in the DMFT derivation is that we consider the limit of high spatial dimensionality (i.e. lattice coordination number $z \to \infty$). To keep the kinetic energy finite, we need to rescale the hopping parameters of the Hamiltonian in Eq. (3.1) as $t_f \to t_f = t_f^*/\sqrt{z}$ (t_f^* is finite). Doing so, the parameter $1/z$ appears as a small parameter in the theory, which is used to control the expansion. We note here that $1/z$ is not a coupling parameter in the original Hamiltonian. Therefore, as mentioned above, this method is suited for the full range of couplings considered. This gives us also a way to estimate the accuracy of our calculations: neglecting terms of order $1/z$ leads to reasonably small errors for the three-dimensional cubic lattice where $z = 6$. DMFT calculations in three dimensions show indeed excellent agreement with QMC calculations [145] and experiments [156].

The first step in this formalism is to derive an effective action of the impurity site (for details see Appendix B) by integrating out the remaining degrees of freedom ($i \neq 0$) in the

3.1 Dynamical Mean-Field Theory (DMFT)

partition function:

$$\frac{1}{Z_{eff}}e^{-S_{eff}} \equiv \frac{1}{Z}\int \prod_{i\neq 0,\sigma} D\tilde{c}^*_{i\sigma}D\tilde{c}_{i\sigma}e^{-S}, \qquad (3.2)$$

where $\tilde{c}_{i\sigma}$, $\tilde{c}^*_{i\sigma}$ are Grassmann variables describing fermions. To leading order in $1/z$ one obtains

$$S_{eff} = -\sum_\sigma \int_0^\beta d\tau_1 \int_0^\beta d\tau_2 \tilde{c}^*_{0\sigma}(\tau_1)\mathcal{G}_\sigma^{-1}(\tau_1-\tau_2)\tilde{c}_{0\sigma}(\tau_2) + U_f\int_0^\beta d\tau \tilde{n}^f_{0\uparrow}(\tau)\tilde{n}^f_{0\downarrow}(\tau). \qquad (3.3)$$

Here we have introduced the *Weiss Green's function*

$$\mathcal{G}_\sigma^{-1}(\tau_1-\tau_2) = -\delta(\tau_1-\tau_2)(\partial_{\tau_2}-\mu_{\sigma f}) - t_f^2\sum_{i,j}{}' G^o_{ij,\sigma}(\tau_1-\tau_2), \qquad (3.4)$$

where $G^o_{ij,\sigma}(\tau_1-\tau_2) = -\langle T\hat{c}_{i\sigma}(\tau_1)\hat{c}^\dagger_{j\sigma}(\tau_2)\rangle^o$ is the interacting Green's function for the fermions, and $\sum_i'$ means summation only over the nearest neighbors of the "impurity site". The expectation values are here calculated in the cavity system without the impurity site, which is indicated by the notation $\langle\ldots\rangle^o$.

The physical content of the Weiss Green's function $\mathcal{G}_\sigma^{-1}(\tau_1-\tau_2)$ is an effective amplitude for fermions to be created on the impurity site at time τ_1, coming from the "external bath" and being destroyed at time τ_2, going back to the "external bath".

In the Matsubara frequency representation the Weiss Green's function has the following form:

$$\mathcal{G}_\sigma^{-1}(i\omega_n) = i\omega_n + \mu_{\sigma f} - t_f^2\sum_{i,j}{}' G^o_{ij,\sigma}(i\omega_n). \qquad (3.5)$$

where $\omega_n = (2n+1)\pi/\beta$ are Matsubara frequencies.

The next step is to express the cavity Green's function by means of the exact Green's function of the original lattice. In the limit of infinite dimensions this relation has the following form:

$$G^o_{ij,\sigma} = G_{ij,\sigma} - \frac{G_{i0,\sigma}G_{0j,\sigma}}{G_{00,\sigma}}. \qquad (3.6)$$

This expression was already derived by Hubbard [157] in his "Hubbard-III" paper in 1964. One can notice that the additional paths, which are contributing in $G_{ij,\sigma}$ and not in $G^o_{ij,\sigma}$ are those which connect site i and j through the impurity site 0. In infinite dimensions one can show [148] that only those paths which go once through the impurity site 0 have to be considered. The contribution of all these additional paths is proportional to $G_{i0,\sigma}G_{0j,\sigma}$, but to count only once the contribution of paths, which are leaving and returning to the impurity site 0, we have to divide this value by $G_{00,\sigma}$.

From Eqs. (3.5) and (3.6) we obtain that, to calculate a new value of the Weiss Green's function, we have to calculate $G_{ij,\sigma} - G_{i0,\sigma}G_{0j,\sigma}/G_{00,\sigma}$. For this purpose we use Fourier transforms of the Green's functions and use the following relationship for the lattice Green's function:

$$G(\mathbf{k}, i\omega_n) = \frac{1}{i\omega_n + \mu_{\sigma f} - \Sigma(i\omega_n) - \varepsilon_{\mathbf{k}}}. \qquad (3.7)$$

Here we would like to mention that we make the assumption, that the self-energy, Σ is a local quantity. This is indeed exact in infinite dimensions [147, 151].

We obtain [148]:

$$\begin{aligned}
t_f^2 G_{ij,\sigma} - \frac{t_f^2 G_{i0,\sigma} G_{0j,\sigma}}{G_{00,\sigma}} &= \int_{-\infty}^{\infty} d\varepsilon \rho(\varepsilon) \frac{\varepsilon^2}{\zeta - \varepsilon} - \left(\int_{-\infty}^{\infty} d\varepsilon \rho(\varepsilon) \frac{\varepsilon}{\zeta - \varepsilon} \right)^2 \bigg/ \int_{-\infty}^{\infty} d\varepsilon \rho(\varepsilon) \frac{1}{\zeta - \varepsilon} \\
&= \zeta - 1 \bigg/ \int_{-\infty}^{\infty} d\varepsilon \rho(\varepsilon) \frac{1}{\zeta - \varepsilon},
\end{aligned} \qquad (3.8)$$

where $\zeta = i\omega_n + \mu - \Sigma(i\omega_n)$. So we get:

$$\mathcal{G}_\sigma^{-1}(i\omega_n) = \Sigma(i\omega_n) + 1 \bigg/ \int_{-\infty}^{\infty} d\varepsilon \rho(\varepsilon) \frac{1}{\zeta - \varepsilon}. \qquad (3.9)$$

And the result which we now obtain is nothing else but the *Dyson equation*:

$$\boxed{\mathcal{G}_\sigma^{-1}(i\omega_n) = \Sigma_\sigma(i\omega_n) + G_\sigma^{-1}(i\omega_n)}. \qquad (3.10)$$

The self-consistency equation for the fermions assumes the simplest form for the *Bethe lattice* which is schematically depicted in Fig. 3.2 and has a semi-elliptic non-interacting density of states

$$\rho(\varepsilon) = \frac{1}{2\pi t_f^{*2}} \sqrt{4t_f^{*2} - \varepsilon^2}. \qquad (3.11)$$

The reason for this simplification is that for the Bethe lattice the summation in Eq. (3.5) is reduced to $i = j$, because all neighbors of the "impurity site" are decoupled from each other. The interacting Green's function for the cavity system is identified with the corresponding expectation value on the impurity site. This means that we identify $G_{ii,\sigma}^o(\tau_1 - \tau_2) = -\langle T \hat{c}_{i\sigma}(\tau_1) \hat{c}_{i\sigma}^\dagger(\tau_2) \rangle^o = -\langle T \hat{c}_{0\sigma}(\tau_1) \hat{c}_{0\sigma}^\dagger(\tau_2) \rangle_0$, where the notation $\langle \ldots \rangle_0$ denotes the expectation value on the impurity site. In passing by, we note that this involves again

3.1 Dynamical Mean-Field Theory (DMFT)

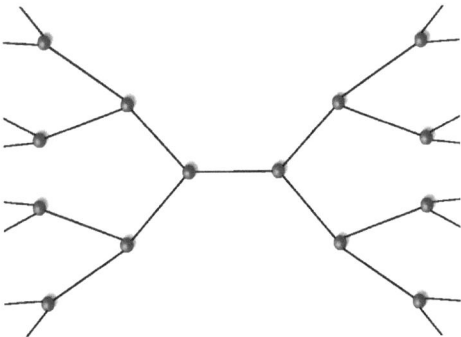

Figure 3.2: Schematic structure of the Bethe lattice (here with coordination number $z = 3$).

an error of order $1/z$ (vanishing in the limit of high dimensionality), since a site at the edge of the cavity has one neighbor less compared to the impurity site. However, in this way we have derived a self-consistency relation, which only involves the impurity site. The self-consistency relation for fermions on the Bethe lattice is therefore

$$\boxed{\mathcal{G}_\sigma^{-1}(i\omega_n) = i\omega_n + \mu_{\sigma f} - t_f^{*2} G_\sigma(i\omega_n)}. \qquad (3.12)$$

As one can see from Fig. 3.2 the Bethe lattice is a bipartite lattice, like the hyper-cubic lattice. From this follows that one can obtain the same type of orders which arise on the hyper-cubic lattice. Therefor, by studying the system on the Bethe lattice, one obtains qualitative insight of the 3D cubic lattice.

The self-consistent DMFT loop has the following structure: we start from an initial guess of the Weiss Green's function. The effective action of the model is then given by Eq. (3.3), which allows us to calculate all local Green's functions and expectation values, including the interacting Green's function. In the case of the Bethe lattice the loop is closed by Eq. (3.12), from which we calculate the new Weiss Green's function, while for a general lattice we use the Dyson equation (3.10). This procedure is repeated until convergence is reached.

The most difficult step in the procedure outlined above is the calculation of the local Green's function from the effective action. Unfortunately one can not do this analytically. So to calculate the Green's function, we return back to the Hamiltonian representation.

Thus one has to find a Hamiltonian which has the same effective action as given by Eq. (3.3). It is easy to see that the corresponding Hamiltonian can not contain only on-site degrees of freedom, because then we would lose retardation effects. The best way to represent the effective action of the (3.3) is via the *Single Impurity Anderson Model* (SIAM). In the SIAM, the impurity site is coupled to a non-interacting fermionic bath which - like the effective action (3.3) - needs to be determined self-consistently in the Dynamical Mean-Field Theory. The SIAM is described by the following Hamiltonian, which allows for a two-sublattice structure:

$$\hat{\mathcal{H}}_{\text{SIAM}} = \sum_{\sigma,\alpha} \left\{ -\mu_{\sigma f}\hat{n}_\alpha^f + U_f \hat{n}_{\uparrow\alpha}^f \hat{n}_{\downarrow\alpha}^f \right\} + \sum_{l,\sigma,\alpha} \left\{ \varepsilon_{l\sigma\alpha} \hat{a}_{l\sigma\alpha}^\dagger \hat{a}_{l\sigma\alpha} + V_{l\sigma\alpha} \left(\hat{f}_{\sigma\alpha}^\dagger \hat{a}_{l\sigma\alpha} + h.c. \right) \right\}. \quad (3.13)$$

Here U_f is the Hubbard on-site interaction between fermions and $\mu_{\sigma f}$ is the chemical potential for spin σ. $\alpha = \pm 1$ is the sublattice index ($\bar{\alpha} = -\alpha$), l labels the noninteracting orbitals of the effective bath, $\varepsilon_{l\sigma\alpha}$ is the energy of the noninteracting orbital l for spin σ on the sub-lattice α and $V_{l\sigma\alpha}$ is the corresponding fermionic hybridization matrix element.

The DMFT calculation is a single site calculation, so to capture phases where size of the unit-cell is more than one lattice site (e. g. antiferromagnetic or charge (particle) density order) we introduce a two sub-lattice structure. In our calculation this corresponds to even and odd DMFT iterations. In general, the number of particles and the correlation functions on different sub-lattices can be different from each other, as they correspond to even and odd DMFT iterations when convergence is reached. The total number of particles in these cases has to be calculated by averaging the number of particles over the last two iterations.

Everything mentioned above was for the non-superconducting case, but this procedure can be easily extended to take into account superconducting long-range order [148, 155]. If one wants to describe superconductivity, in addition to the normal Green's function, one also has to introduce the superconducting Green's function $F(\mathbf{k}, \tau) = -\langle T \hat{c}_{\mathbf{k}\uparrow}(\tau) \hat{c}_{-\mathbf{k}\downarrow}(0) \rangle$. For non-zero F, it is better to work with Nambu spinors $\hat{\Psi}^\dagger(\mathbf{k}, \tau) = (\hat{c}_{\mathbf{k},\uparrow}^\dagger, \hat{c}_{-\mathbf{k},\downarrow})$ and with the matrix formulation of the one particle Green's function:

$$\hat{G}(\mathbf{k}, \tau) = -\langle T \hat{\Psi}_\mathbf{k}(\tau) \hat{\Psi}_\mathbf{k}^\dagger(0) \rangle = \begin{pmatrix} G_\uparrow(\mathbf{k}, \tau) & F(\mathbf{k}, \tau) \\ F^\star(\mathbf{k}, \tau) & -G_\downarrow(-\mathbf{k}, -\tau) \end{pmatrix}. \quad (3.14)$$

Following the cavity method, we will obtain the same type of equations as for the non-superconducting case, but instead of usual functions one will have matrices. So the Dyson equation will have the following form:

$$\hat{\mathcal{G}}^{-1}(i\omega_n) = \hat{\Sigma}(i\omega_n) + \hat{G}^{-1}(i\omega_n), \tag{3.15}$$

where $\hat{G}(\omega)$ is the matrix of interacting Green's functions, $\hat{\Sigma}(\omega)$ is the self-energy matrix and $\hat{\mathcal{G}}$ is the matrix of Weiss Green's functions.

The self-consistency equation for the Bethe lattice in the matrix representation has the following form:

$$\hat{\mathcal{G}}^{-1}(i\omega_n) = (i\omega_n + \frac{1}{2}\Delta\mu)\hat{I} + \bar{\mu}_f\hat{\sigma}_z - t_f^{*2}\hat{\sigma}_z\hat{G}_\sigma(i\omega_n)\hat{\sigma}_z, \tag{3.16}$$

where $\hat{I}$ is the unit matrix, $\hat{\sigma}_z$ is the z Pauli matrix, $\bar{\mu}_f = (\mu_\uparrow + \mu_\downarrow)/2$ and $\Delta\mu = \mu_\uparrow - \mu_\downarrow$.

To describe superconductivity one has to add an additional term to the SIAM Hamiltonian:

$$\begin{aligned}\hat{\mathcal{H}}_{\text{SIAM}} &= \sum_{\alpha=\pm 1}\left\{-\mu_{\sigma f}\hat{n}^f_\alpha + U_f\hat{n}^f_{\uparrow\alpha}\hat{n}^f_{\downarrow\alpha}\right\} + \sum_{l,\sigma}\left\{\varepsilon_{l\sigma\alpha}\hat{a}^\dagger_{l\sigma\alpha}\hat{a}_{l\sigma\alpha} + V_{l\sigma\alpha}\left(\hat{f}^\dagger_{\sigma\alpha}\hat{a}_{l\sigma\alpha} + h.c.\right)\right. \\ &\quad \left. + W_{l\alpha}\left(\hat{a}^\dagger_{l\uparrow\alpha}\hat{a}^\dagger_{l\downarrow\alpha} + h.c.\right)\right\}. \end{aligned} \tag{3.17}$$

As there is no analytical solution of the SIAM, one has to solve it numerically. For this purpose there exist different impurity solvers. In section 3.4 we will discuss some of them.

In the end of this section we would like to mention that it is clear that the final result of the DMFT calculations should not depend on the initial conditions of the self-consistency loop. However, for physical reasons it can happen that the self-consistent DMFT procedure yields multiple stable solutions. To find the ground state of the system in those cases, we need to compare the energies of the coexisting solutions. The ground state will correspond to the solution with the lowest energy. For this purpose we need to calculate the total energy which is given as follows:

$$\frac{E}{N} = \frac{\mathcal{E}_{kin}}{N} + \frac{\mathcal{E}_{int}}{N}, \tag{3.18}$$

where $\frac{\mathcal{E}_{int}}{N}$ depends on the type of interactions on the impurity site. For the Hubbard model it is given by the following equation

$$\frac{\mathcal{E}_{int}}{N} = \frac{U_f}{2}\left(\langle \hat{n}^f_{\uparrow,-1}\hat{n}^f_{\downarrow,-1}\rangle + \langle \hat{n}^f_{\uparrow,1}\hat{n}^f_{\downarrow,1}\rangle\right), \tag{3.19}$$

where ± 1 are sublattice indices.

More difficult is to calculate the kinetic energy. For the para- and ferro-magnetic phases, the kinetic energy is given by the following equation:

$$\frac{\mathcal{E}_{kin}}{N} = k_B T \sum_{n,\sigma} \int_{-\infty}^{\infty} d\varepsilon \, \varepsilon \rho(\varepsilon) G_\sigma(\varepsilon, i\omega_n), \tag{3.20}$$

where $G_\sigma(\varepsilon, i\omega_n) = G_\sigma(\mathbf{k}, i\omega_n)$ is given by Eq. (3.7), while for a system, which has antiferromagnetic or charge density order (fermionic density oscillation), the kinetic energy is given as follows (for detail see Appendix E):

$$\frac{\mathcal{E}_{kin}}{N} = k_B T \sum_{n,\sigma} \int_{-\infty}^{\infty} d\varepsilon \, \varepsilon \rho(\varepsilon) \mathcal{B}_\sigma(\varepsilon, i\omega_n), \tag{3.21}$$

where

$$\mathcal{B}_\sigma(\varepsilon, i\omega_n) = \frac{1}{\sqrt{\zeta_{\sigma,1} \zeta_{\sigma,-1}} - \varepsilon} \tag{3.22}$$

and $\zeta_{\sigma\alpha} = \omega + \mu_{\sigma f} - \Sigma_{\sigma\alpha}(\omega)$.

3.2 Generalized Dynamical Mean-Field Theory (GDMFT)

3.2.1 Method

The dynamical mean-field theory (DMFT) was invented to study the behavior of correlated fermions. Nowadays ultracold atomic gases give us the possibility to investigate mixtures of fermions and bosons in optical lattices. These systems are well described by the Bose-Fermi Hubbard model:

$$\begin{aligned}\hat{\mathcal{H}} &= -t_f \sum_{\langle i,j \rangle, \sigma} \hat{c}_{i\sigma}^\dagger \hat{c}_{j\sigma} - t_b \sum_{\langle i,j \rangle} \hat{b}_i^\dagger \hat{b}_j - \sum_{i,\sigma} \mu_{\sigma f} \hat{n}_i^f - \mu_b \sum_i \hat{n}_i^b \\ &+ U_f \sum_i \hat{n}_{i\uparrow}^f \hat{n}_{i\downarrow}^f + \frac{U_b}{2} \sum_i \hat{n}_i^b (\hat{n}_i^b - 1) + U_{fb} \sum_i \hat{n}_i^f \hat{n}_i^b + g \sum_i \left(\hat{b}_i^\dagger \hat{c}_{i\uparrow} \hat{c}_{i\downarrow} + h.c \right), \end{aligned} \tag{3.23}$$

3.2 Generalized Dynamical Mean-Field Theory (GDMFT)

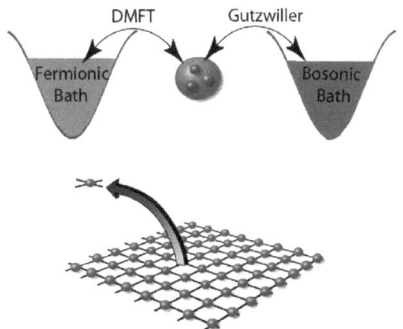

Figure 3.3: Schematic picture of the Generalized Dynamical Mean-Field theory (GDMFT): within the GDMFT approach the full many-body lattice problem is replaced by a single-site problem, which is coupled to the fermionic bath as in the "usual" DMFT and to the bosonic bath via the Gutzwiller approach.

where $\hat{c}_{i\sigma}^\dagger (\hat{b}_i^\dagger)$ is the fermionic (bosonic) creation operator at site i, while $\hat{n}_{i\sigma}^f = \hat{c}_{i\sigma}^\dagger \hat{c}_{i\sigma}$ and $\hat{n}_i^b = \hat{b}_i^\dagger \hat{b}_i$ are the number operators for fermions and bosons at site i respectively and $\hat{n}_i^f = \hat{n}_{i\uparrow}^f + \hat{n}_{i\downarrow}^f$. μ_b and $\mu_{\sigma f}$ are the chemical potentials for boson and fermions with spin σ. U_f, U_b and U_{fb} are local interactions between fermions, bosons and fermions and bosons respectively. $\langle i,j \rangle$ denotes summation over nearest neighbors and $t_{f(b)}$ is the tunneling amplitude for fermions (bosons). g is the Feshbach term, which describes processes where two fermions on the same site combine to form a bosonic molecule.

To describe such a system one has to *generalize* the DMFT method and include also bosonic degrees of freedom [20, 21]. In addition to rescaling fermionic hopping parameters, we also have to rescale the bosonic hopping parameters of the Hamiltonian Eq. (3.23) according to $t_b \to t_b = t_b^*/z$ to keep the kinetic energy finite (t_b^* is finite).

The first step in this formalism, as well as in the "usual" DMFT, is to derive the effective action of the impurity site (for details see Appendix B) by integrating out the remaining degrees of freedom ($i \neq 0$) in the partition function:

$$\frac{1}{Z_{eff}} e^{-S_{eff}} \equiv \frac{1}{Z} \int \prod_{i \neq 0,\sigma} D\tilde{c}_{i\sigma}^* D\tilde{c}_{i\sigma} D\tilde{b}_i^* D\tilde{b}_i e^{-S}, \qquad (3.24)$$

where $\tilde{c}_{i\sigma}$, $\tilde{c}_{i\sigma}^*$ are Grassmann variables describing fermions and $\tilde{b}_i$, $\tilde{b}_i^*$ are $\mathbb{C}$-numbers describing bosons. To leading order in $1/z$ one obtains

$$\begin{aligned}
S_{eff} &= -\sum_\sigma \int_0^\beta d\tau_1 \int_0^\beta d\tau_2 \tilde{c}_{0\sigma}^\star(\tau_1) \mathcal{G}_\sigma^{-1}(\tau_1 - \tau_2) \tilde{c}_{0\sigma}(\tau_2) + \int_0^\beta d\tau\, \tilde{b}_0^\star(\tau)(\partial_\tau - \mu_b)\tilde{b}_0(\tau) \\
&\quad - t_b \int_0^\beta d\tau {\sum_i}' (\Phi_i^o(\tau)\tilde{b}_0^\star(\tau) + c.c) + U_f \int_0^\beta d\tau\, \tilde{n}_{0\uparrow}^f(\tau)\tilde{n}_{0\downarrow}^f(\tau) + U_{fb} \int_0^\beta d\tau\, \tilde{n}_0^f(\tau)\tilde{n}_0^b(\tau) \\
&\quad + U_b \int_0^\beta d\tau\, \tilde{n}_0^b(\tau)(\tilde{n}_0^b(\tau) - 1) + g \int_0^\beta d\tau (\tilde{c}_{0\downarrow}^\star(\tau) \tilde{c}_{0\uparrow}^\star(\tau) \tilde{b}_0(\tau) + c.c)\,.
\end{aligned} \quad (3.25)$$

Here $\Phi_i^o(\tau) = \langle \hat{b} \rangle^o$ is the bosonic superfluid parameter, which is static. $\mathcal{G}_\sigma^{-1}(\tau_1 - \tau_2)$ is the Weiss Green's function, which we already introduced in the previous section Eq. (3.4).

The next step in the derivation is that the expectation values in the cavity system are identified with the expectation values on the impurity site. This means that we identify $\Phi_i^o(\tau) = \langle \hat{b} \rangle^o = \langle \hat{b} \rangle_0$ and $G_{ii,\sigma}^o(\tau_1 - \tau_2) = -\langle T \hat{c}_{i\sigma}(\tau_1)\hat{c}_{i\sigma}^\dagger(\tau_2)\rangle^o = -\langle T \hat{c}_{0\sigma}(\tau_1)\hat{c}_{0\sigma}^\dagger(\tau_2)\rangle_0$, where the notation $\langle \ldots \rangle_0$ means expectation value for the impurity site. However, in this way, we have derived a self-consistency relation, which only involves the impurity site.

By inspecting these self-consistency relations, it becomes clear that the bosonic part corresponds to the Gutzwiller approximation [158–160], whereas the fermionic part corresponds to the DMFT [147–155]. The two are coupled by the on-site density-density interaction. We remark here that this derivation shows that the Gutzwiller approximation for bosons is exact in infinite dimensions, and, like the DMFT, valid for arbitrary couplings in the Hamiltonian. Therefore, this approximation is able to describe the whole phase-diagram, in particular, the transition from superfluid to Mott-insulator. This point is not always appreciated in the literature, where the Gutzwiller approximation is sometimes regarded as a strong-coupling approximation.

Summarizing, the GDMFT employed in the calculations in this thesis consists of the DMFT algorithm for the fermions, combined with bosonic Gutzwiller mean-field theory. The bosons are described by the superfluid order parameter $\Phi_i^o(\tau) = \langle \hat{b}(\tau) \rangle$, while the fermions are characterized by the Weiss Green's function $\mathcal{G}_\sigma^{-1}(i\omega_n)$, which is given by Eq. (3.5). Schematically the GDMFT is depicted in Fig. 3.3.

The self-consistent GDMFT loop has a similar structure as in the case of DMFT: We start from an initial guess of the Weiss Green's function and the superfluid order parameter. The effective action of the model is then given by Eq. (3.25), which allows us to calculate all local Green's functions and expectation values, including the interacting Green's functions and the superfluid order parameter. The loop is closed by Eq. (3.5), from which we calculate the new Weiss Green's function. This procedure is repeated until convergence is

3.2 Generalized Dynamical Mean-Field Theory (GDMFT)

reached.

As we already mentioned in the previous section, the DMFT loop has the simplest form for the Bethe lattice. In this case, the Weiss Green's function is calculated by Eq. (3.16), while for general lattice one has to use the Dyson equation (3.15). For this purpose, one has to calculate the self-energy. As we show in the appendix D, one can express the self-energy via different Green's functions:

$$\Sigma_\sigma(i\omega_n) = \frac{\left(U_f Q_{ff\sigma}(i\omega_n) + U_{fb} Q_{fb\sigma}(i\omega_n) + \sigma g Q^\star_{g\bar{\sigma}\sigma}(i\omega_n)\right) G^\star_{\bar{\sigma}}(i\omega_n)}{G_\sigma(i\omega_n) G^\star_{\bar{\sigma}}(i\omega_n) + F(\sigma i\omega_n) F^\star(\bar{\sigma} i\omega_n)}$$
$$+ \frac{\left(\sigma U_f Q_{ff,\sigma\bar{\sigma}}(i\omega_n) + \sigma U_{fb} Q_{fb\sigma\bar{\sigma}}(i\omega_n) + g Q^\star_{g\bar{\sigma}}(i\omega_n)\right) F^\star(\bar{\sigma} i\omega_n)}{G_\sigma(i\omega_n) G^\star_{\bar{\sigma}}(i\omega_n) + F(\sigma i\omega_n) F^\star(\bar{\sigma} i\omega_n)}, \quad (3.26)$$

$$\Sigma_{SC}(i\omega_n) = \frac{\left(U_f Q_{ff\uparrow}(i\omega_n) + U_{fb} Q_{fb\uparrow}(i\omega_n) + g Q^\star_{g\downarrow\uparrow}(i\omega_n)\right) F(i\omega_n)}{G_\uparrow(i\omega_n) G^\star_\downarrow(i\omega_n) + F(i\omega_n) F^\star(-i\omega_n)}$$
$$- \frac{\left(U_f Q_{ff,\uparrow\downarrow}(i\omega_n) + U_{fb} Q_{fb\uparrow\downarrow}(i\omega_n) + g Q^\star_{g\downarrow}(i\omega_n)\right) G_\uparrow(i\omega_n)}{G_\uparrow(i\omega_n) G^\star_\downarrow(i\omega_n) + F(i\omega_n) F^\star(-i\omega_n)}, \quad (3.27)$$

where

$$Q_{ff\sigma}(i\omega_n) = \langle\langle \hat{f}_\sigma \hat{f}^\dagger_{\bar{\sigma}} \hat{f}_{\bar{\sigma}}, \hat{f}^\dagger_\sigma \rangle\rangle_\omega, \qquad Q_{ff\sigma\bar{\sigma}}(i\omega_n) = \langle\langle \hat{f}_\sigma \hat{f}^\dagger_{\bar{\sigma}} \hat{f}_{\bar{\sigma}}, \hat{f}_{\bar{\sigma}} \rangle\rangle_\omega,$$
$$Q_{fb\sigma}(i\omega_n) = \langle\langle \hat{f}_\sigma \hat{b}^\dagger \hat{b}, \hat{f}^\dagger_\sigma \rangle\rangle_\omega, \qquad Q_{fb\sigma\bar{\sigma}}(i\omega_n) = \langle\langle \hat{f}_\sigma \hat{b}^\dagger \hat{b}, \hat{f}_{\bar{\sigma}} \rangle\rangle_\omega, \quad (3.28)$$
$$Q_{g\sigma}(i\omega_n) = \langle\langle \hat{f}_\sigma \hat{b}^\dagger, \hat{f}^\dagger_\sigma \rangle\rangle_\omega, \qquad Q_{yv\bar{\sigma}}(i\omega_n) = \langle\langle \hat{f}_\sigma \hat{b}^\dagger, \hat{f}_{\bar{\sigma}} \rangle\rangle_\omega$$

and $\langle\langle \hat{A}, \hat{B} \rangle\rangle = \int_0^\beta d\tau e^{i\omega_n \tau} \langle \hat{A}(\tau) \hat{B}(0) \rangle$.

As in the DMFT, to solve effective action we are returning back to the Hamiltonian representation. The best way to represent effective action (3.25) is via a *Generalized Single Impurity Anderson Model* (GSIAM). As in the conventional Single Impurity Anderson model (SIAM), the impurity site is coupled to a non-interacting fermionic bath which needs to be determined self-consistently in the Dynamical Mean-Field Theory. In addition, the GSIAM now also contains a bosonic degree of freedom on the "impurity site", which is self-consistently coupled to the superfluid order parameter, according to the Gutzwiller mean-field theory. In summary, the GSIAM is described by the following Hamiltonian, which allows for a two-sublattice structure:

$$\hat{\mathcal{H}}_{\text{GSIAM}} = \sum_{\alpha=\pm 1} \left[\hat{\mathcal{H}}_b^\alpha + \hat{\mathcal{H}}_{fb}^\alpha + \hat{\mathcal{H}}_f^\alpha \right], \tag{3.29}$$

$$\hat{\mathcal{H}}_b^\alpha = -zt_b(\varphi_{\bar{\alpha}} \hat{b}_\alpha^\dagger + \varphi_{\bar{\alpha}}^\star \hat{b}_\alpha) + \frac{U_b}{2} \hat{n}_\alpha^b (\hat{n}_\alpha^b - 1) - \mu_b \hat{n}_\alpha^b, \tag{3.30}$$

$$\hat{\mathcal{H}}_{fb}^\alpha = U_{fb} \hat{n}_\alpha^f \hat{n}_\alpha^b + g \left(\hat{c}_{\downarrow\alpha}^\dagger \hat{c}_{\uparrow\alpha}^\dagger \hat{b}_\alpha + h.c. \right), \tag{3.31}$$

$$\hat{\mathcal{H}}_f^\alpha = -\mu_{\sigma f} \hat{n}_\alpha^f + U_f \hat{n}_{\uparrow\alpha}^f \hat{n}_{\downarrow\alpha}^f + \tag{3.32}$$
$$+ \sum_{l,\sigma} \left\{ \varepsilon_{l\sigma\alpha} \hat{a}_{l\sigma\alpha}^\dagger \hat{a}_{l\sigma\alpha} + V_{l\sigma\alpha} \left(\hat{f}_{\sigma\alpha}^\dagger \hat{a}_{l\sigma\alpha} + h.c. \right) + W_{l\alpha} \left(\hat{a}_{\uparrow\alpha}^\dagger \hat{a}_{l\downarrow\alpha}^\dagger + h.c. \right) \right\},$$

where U_f, U_b, and U_{fb} are on-site Hubbard interactions between fermions, bosons, and fermions and bosons respectively. $\mu_{\sigma f}$ and μ_b are the chemical potentials for fermions and bosons. g is the Feshbach term. $\alpha = \pm 1$ is the sublattice index ($\bar{\alpha} = -\alpha$), z is the lattice coordination number, $\varphi_\alpha = \langle \hat{b}_\alpha \rangle$ is the superfluid order parameter on sublattice α. l labels the noninteracting orbitals of the effective bath, $\varepsilon_{l\sigma\alpha}$ is the energy of the noninteracting orbital l for spin σ on the sub-lattice α and $V_{l\sigma\alpha}$ is the corresponding fermionic hybridization matrix element. $W_{l\alpha}$ is a local superconducting amplitude of the noninteracting orbital l on the sub-lattice α.

3.2.2 Energy Calculations

Now we would like again to consider the case when the self-consistent GDMFT procedure yields multiple stable solutions. To find the ground state of the system in those cases, we need to compare the energies of the coexisting solutions. The ground state will correspond to the solution with the lowest energy. For this purpose we need to calculate the total energy which is given as follows:

$$\frac{E}{N} = \frac{\mathcal{E}_{kin}}{N} + \frac{\mathcal{E}_{int}}{N}, \tag{3.33}$$

where

$$\frac{\mathcal{E}_{int}}{N} = \frac{1}{2} \sum_{\alpha=\pm 1} \left(U_{fb} \langle \hat{n}_\alpha^f \hat{n}_\alpha^b \rangle + U_f \langle \hat{n}_{\uparrow\alpha}^f \hat{n}_{\downarrow\alpha}^f \rangle + \frac{U_b}{2} \langle \hat{n}_\alpha^b (\hat{n}_\alpha^b - 1) \rangle + g(\langle \hat{c}_{\downarrow\alpha}^\dagger \hat{c}_{\uparrow\alpha}^\dagger \hat{b} \rangle + c.c) \right) \tag{3.34}$$

3.2 Generalized Dynamical Mean-Field Theory (GDMFT)

and $\alpha = \pm 1$ are sublattice indices. The kinetic energy is given by the following equation:

$$\frac{\mathcal{E}_{kin}}{N} = -zt_b\varphi_{-1}\varphi_1 + \frac{\mathcal{E}_{kin}^f}{N}, \qquad (3.35)$$

where $\mathcal{E}_{kin}^f$ is calculated according to Eq. (3.21).

3.2.3 Summary

We close this section with a short summary of the method. The GDMFT technique is a combination of the DMFT and Gutzwiller approaches. We have shown that it is exact in infinite dimensions, and it is assumed to be a good approximation for three spatial dimensions. The only small parameter in this method is $1/z$ (where z is the lattice coordination number). The GDMFT, therefore, incorporates local correlations between bosons and fermions in a fully non-perturbative fashion. Non-local correlations, on the other hand, can be calculated only on a mean-field level.

Since the fermions are treated with a *dynamical* mean-field, their quantum fluctuations are also captured. Higher orders in $1/z$ could make quantitative changes, but no qualitative changes are expected. The bosons on the other hand are treated within static mean field theory and couple only to the bosonic order parameter. Although this is indeed exact in infinite dimensions, for a finite number of spatial dimensions even normal (i.e. non-superfluid) bosons will hop. This will e.g. effect the fluctuations in the boson number $\langle \hat{n}_b^2 \rangle - \langle \hat{n}_b \rangle^2$. Within the Gutzwiller approximation this quantity is zero in the Mott insulator and the alternating Mott insulator phase (which will be defined later). Inclusion of normal hopping would lead to finite fluctuations. This effect is however not essential for the physics considered in this thesis. In future calculations, normal bosonic hopping could be included via the recently developed Bosonic DMFT (BDMFT) [161, 162].

The above derivation is valid independently of temperature and impurity solver. Therefore, GDMFT also gives a reliable description of Bose-Fermi mixtures in an optical lattice at any finite temperature.

3.3 Real-Space Dynamical Mean-Field Theory (R-DMFT)

The methods described in this chapter so far are for homogeneous systems. However, experimentally the spatial inhomogeneity due to the harmonic confinement potential is always present, leading to a spatially varying local density. In this case the concept of long range order is questionable and ordered phases are expected to develop on finite length scales. To study such systems we develop a *real-space* extension of the *dynamical mean-field theory* [163]. Within this *real space dynamical mean-field theory* (R-DMFT) the self-energy is taken to be local, which is exact in the infinite dimensional limit [147, 148, 151]. However, in an inhomogeneous system it depends on the lattice site, i.e. $\Sigma_{ij\sigma} = \Sigma_\sigma^{(i)} \delta_{ij}$, where δ_{ij} is a Kronecker delta. Formerly, a similar scheme has been developed for systems with inhomogeneity in one direction [164]. Only recently, problems with full inhomogeneity have been investigated, in particular the Falicov-Kimball model [165, 166], disordered systems [167] and paramagnetic states of cold fermionic atoms [168, 169].

Repulsively interacting fermions in an optical lattice almost perfectly implement the Hubbard Hamiltonian

$$\hat{\mathcal{H}} = -t_f \sum_{\langle ij \rangle, \sigma} \hat{c}_{i\sigma}^\dagger \hat{c}_{j\sigma} + U_f \sum_i \hat{n}_{i\uparrow} \hat{n}_{i\downarrow} + \sum_{i\sigma} (V_i - \mu_\sigma) \hat{n}_{i\sigma}, \tag{3.36}$$

where $\hat{n}_{i\sigma} = \hat{c}_{i\sigma}^\dagger \hat{c}_{i\sigma}$, and $\hat{c}_{i\sigma}$ ($\hat{c}_{i\sigma}^\dagger$) are fermionic annihilation (creation) operators for an atom with spin σ at site i, t_f is the hopping amplitude between nearest neighbor sites $\langle ij \rangle$, U_f is the on-site interaction, μ_σ is the (spin-dependent) chemical potential and $V_i = V_0 r_i^2$ is the harmonic confinement potential. The parameters of this model are, as we described in section 2.5, tunable in experiments by changing the lattice amplitude and via Feshbach resonances.

Within the R-DMFT method, the Hamiltonian is mapped onto a set of single site problems. The physics of the lattice site i is described by the local effective action [148]

$$\begin{aligned} S_{\text{eff}}^{(i)} &= -\int d\tau \int d\tau' \sum_\sigma \tilde{c}_{i\sigma}^*(\tau) \mathcal{G}_0^{(i)}(\sigma, \tau - \tau')^{-1} \tilde{c}_{i\sigma}(\tau') \\ &\quad - U_f \int d\tau \tilde{n}_{i\uparrow}(\tau) \tilde{n}_{i\downarrow}(\tau), \end{aligned} \tag{3.37}$$

which explicitly depends on the site index i. The Weiss Green's function $\mathcal{G}_0^{(i)}(\sigma, \tau - \tau')$ as in the "usual" dynamical mean-field theory (DMFT) simulates the effect of all other sites. The difference is that in the R-DMFT, we define a Weiss Green's function $\mathcal{G}_0^{(i)}(\sigma, \tau - \tau')$

3.3 Real-Space Dynamical Mean-Field Theory (R-DMFT)

for each lattice site.

The self-consistency loop for the R-DMFT has the following structure: we start from an initial guess of a set of a Weiss Green's functions $\{\mathcal{G}_0^{(i)}(\sigma, \tau - \tau')\}$. After solving the action (3.37), we can calculate a set of the local self-energies $\{\Sigma^{(i)}(\sigma, i\omega_n)\}$. The next step is to determine the interacting lattice Green's function from the Dyson equation in real-space representation

$$\mathbf{G}(\sigma, i\omega_n)^{-1} = \mathbf{G}_0(\sigma, i\omega_n)^{-1} - \mathbf{\Sigma}(\sigma, i\omega_n), \tag{3.38}$$

where the boldface notation indicates that the quantities are matrices labeled by site indices i and j and ω_n are the Matsubara frequencies. The size of these matrices are $N \times N$, where N is the number of the lattice sites. The non-interacting lattice Green's function is given by

$$\mathbf{G}_0(\sigma, i\omega_n)^{-1} = (\mu_\sigma + i\omega_n)\mathbf{1} - \mathbf{t} - \mathbf{V}, \tag{3.39}$$

where $\mathbf{1}$ is the unit matrix. The matrix elements t_{ij} are hopping amplitudes for a given lattice structure, i.e. $t_{ij} \neq 0$ if fermions jump from site i to site j or vice versa. $V_{ij} = \delta_{ij} V_i$ represents a spatially varying potential. The diagonal elements of the lattice Green's function are nothing else but the interacting local Green's functions, i.e. $G^{(i)}(\sigma, i\omega_n) = G_{ii}(\sigma, i\omega_n)$. Finally, the Weiss mean-field is obtained from the local Dyson equation

$$\mathcal{G}_0^{(i)}(\sigma, i\omega_n)^{-1} = G^{(i)}(\sigma, i\omega_n)^{-1} + \Sigma^{(i)}(\sigma, i\omega_n), \tag{3.40}$$

which closes the set of self-consistency equations. In Figure 3.4 we schematically depicted the R-DMFT loop.

The most difficult step in this procedure is calculating the local action (3.37). This step is, however, similar to the solution of the local action in a homogeneous DMFT calculation. The difference is that in the present case the Weiss field $\mathcal{G}_0^{(i)}(\sigma, \tau)$ is obtained via the Real-Space Dyson equation (3.38), which incorporates the effect of the spatial inhomogeneity. This implies that for the numerical solution of the local action we can use standard techniques, which have proved to be reliable and efficient.

In practice the self-consistent solution is obtained iteratively from the initial Weiss mean-fields $\mathcal{G}_0^{(i)}(\sigma, i\omega_n)$ which are chosen differently for different spin σ and lattice sites i. Then the solutions with staggered magnetization or phase separation are obtained naturally in contrast to the standard DMFT, where an additional sublattice structure has to be added [148].

The R-DMFT scheme can also be used to study Bose-Fermi mixtures in a harmonic

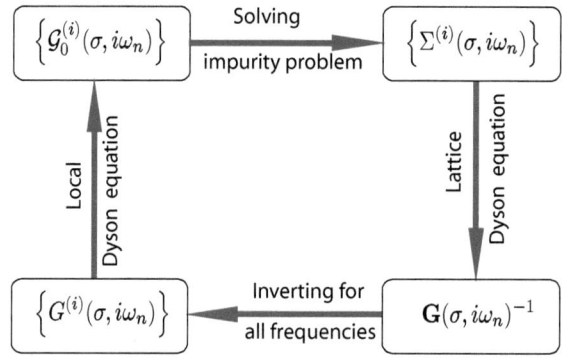

Figure 3.4: Schematic picture of the Real-space Dynamical Mean-Field theory (R-DMFT) loop.

trap. As we mentioned in the previous section, for characterizing the Bose-Fermi mixture we need in addition to the Weiss Green's function, which describes fermions, also the superfluid order parameter which describes bosons. One can calculate new values for Weiss Green's functions as described above, while for calculating the new values for the superfluid order parameter one have to average the superfluid order parameters of the neighboring sites.

We would like to mention that the R-DMFT calculations are much more demanding compared to the "usual" DMFT calculations. But within the R-DMFT significantly larger systems can be investigated than those studied by quantum Monte Carlo [144–146], for which only homogeneous data are available in two and three dimensions. The computational effort scales polynomially with the number of lattice sites N within the R-DMFT. The application of the real-space Dyson equation requires a sparse matrix inversion for each frequency, which scales as $N^{3/2}$. The number of impurity solver calculations per R-DMFT-run is linear in N, but can be kept small due to symmetries. Moreover, the solution of the real-space Dyson equation can be parallelized over the frequencies and the impurity solver calculations can be parallelized over the lattice sites.

3.4 Impurity Solvers

As we have already shown, to solve a DMFT loop, one has to solve the Anderson impurity model. Due to the fact that there is no analytical solution available, one has to solve it numerically. There exist several numerical methods to solve this problem. Among them the most powerful ones are the *Numerical renormalization Group* (NRG) [170–178], *Exact Diagonalization* (ED) [148, 179–181] and *Quantum Monte Carlo* (QMC) [148, 182–185]. In this thesis we are using only NRG and ED as impurity solvers. In the next subsections we will consider these two method in detail.

3.4.1 Exact Diagonalization (ED)

Exact Diagonalization (ED)[148, 179–181] as an impurity solver, was introduced by Caffarel and Krauth [179]. The first step in this algorithm is truncation of the infinite number of orbitals in the single impurity Anderson model (SIAM) (Eq. 3.17 for pure fermions and Eq. (3.29) for Bose-Fermi mixtures) and considering a finite (and relatively small) number n_s of orbitals. The resulting finite-size problem is different from the finite-size problem of a finite number of lattice sites of the original Hubbard model and the truncation procedure can be viewed as using a finite number of parameters (energy scales) to describe the local dynamics encoded in the Weiss Green's function (which we derive in the Appendix D):

$$\mathcal{G}^{-1}_{\sigma,SIAM}(i\omega_n) = i\omega_n + \mu_\sigma + \sum_{l=1}^{n_s} V_{l\sigma}^2 \frac{i\omega_n + \varepsilon_{l\bar{\sigma}}}{(\varepsilon_{l\sigma} - i\omega_n)(\varepsilon_{l\bar{\sigma}} + i\omega_n) + W_l^2}, \quad (3.41)$$

$$\mathcal{F}^{-1}_{\sigma,SIAM}(i\omega_n) = -\sum_{l=1}^{n_s} \frac{V_{l\uparrow} V_{l\downarrow} W_l}{(\varepsilon_{l\sigma} - i\omega_n)(\varepsilon_{l\bar{\sigma}} + i\omega_n) + W_l^2}, \quad (3.42)$$

where ω_n are the Matsubara frequencies.

After truncation the Hilbert space of the SIAM only contains a finite number of states: 4^{n_s+1} for two component fermions, $(N_b + 1)2^{n_s+1}$ for the mixture of spinless fermions and bosons and $(N_b + 1)4^{n_s+1}$ for the mixture of the two component fermions and bosons, where N_b is the bosonic cut-off. As we deal with a finite size system we can diagonalize the Hamiltonian and find the eigenvalues E_i and eigenvectors $|i\rangle$ of the problem. Knowing the eigenvalues and eigenvectors we can calculate different Green's functions:

$$G_{AB}(i\omega_n) = \langle\langle \hat{A}, \hat{B} \rangle\rangle_\omega = -\frac{1}{Z} \sum_{n,m} \langle n|\hat{A}|m\rangle \langle m|\hat{B}|n\rangle \frac{e^{-\beta E_n} + e^{-\beta E_m}}{E_m - E_n - i\omega_n}, \quad (3.43)$$

where β is the inverse temperature and

$$Z = \sum_n e^{-\beta E_n} \qquad (3.44)$$

is the partition function.

To speed up the diagonalization process one can use the block-diagonal structure of the Hamiltonian. Depending on the problem, the Hamiltonian conserves the total number of fermions N_f and/or the total magnetization M_z. When both of these quantum numbers are conserved, the size of each block in the Hamiltonian matrix is much smaller, compared to the case when only one of the quantum numbers (N_f, M_z) is conserved. So in this case one can consider more orbitals.

For the Bethe lattice, knowing the interacting Green's function $G_\sigma(i\omega_n) = \langle\langle \hat{f}_\sigma, \hat{f}_\sigma^\dagger \rangle\rangle_\omega$, and the superconducting Green's function $F(i\omega_n) = \langle\langle \hat{f}_\uparrow, \hat{f}_\downarrow \rangle\rangle_\omega$ we can calculate the Weiss Green's function using equation (3.16), while for more complicated lattices the Weiss Green's function can be calculated using the Dyson equation (3.15). For this purpose we have to calculate the self-energy using Eqs. (3.26) and (3.27).

The next step is to determine new parameters for the SIAM Hamiltonian. We therefore compare the Weiss functions calculated from (3.41) and (3.42) to the ones calculated from exact diagonalization:

$$\chi_{\alpha\beta} = \frac{1}{2(N_{max}+1)} \sum_{n=0}^{N_{max}} \left(\sum_\sigma |\mathcal{G}^\alpha_{\sigma,SIAM}(i\omega_n) - \mathcal{G}^\alpha_{\sigma,ex}(i\omega_n)|^\beta + 2|\mathcal{F}^\alpha_{SIAM}(i\omega_n) - \mathcal{F}^\alpha_{ex}(i\omega_n)|^\beta \right), \qquad (3.45)$$

where N_{max} is the number of the different Matsubara frequency, $\alpha = \pm 1$ and $\beta = 1, 2$. For different set of α and β convergence speed could be different, so changing these parameters one can choose those, which are optimal for the problem.

The minimization process works as follows: we start from an initial guess of the SIAM parameters ($\epsilon_{l\sigma}$, $V_{l\sigma}$ and W_l), and then using the Green's function calculated by Eq. (3.43) we calculate the Weiss Green's functions $\mathcal{G}^\alpha_{\sigma,ex}(i\omega_n)$ and $\mathcal{F}^\alpha_{\sigma,ex}(i\omega_n)$. The next step is to fit the Weiss Green's functions calculated by Eqs. (3.41) and (3.42) and find a new set of parameters for the SIAM. After this step we start everything from the beginning with the new set of parameters. This procedure is repeated until convergence is reached.

Here we would like to mention that exact diagonalization (ED) is a non-perturbative impurity solver and describes the system for all parameter ranges from weak to strong coupling. Using ED one can also perform calculations for zero temperature as well as for finite-temperature. The disadvantage of ED is that it does not describe low-frequency spectral function with high accuracy. The reason is that since we restrict to a finite number of orbitals (n_s order of one), the number of peaks in the spectral function are limited.

In the end of this sub-section we also consider how one can use ED as an impurity solver for the R-DMFT method. In this case we start from an initial guess of the SIAM parameters $\{\varepsilon_{l\sigma}^{(i)}, V_{l\sigma}^{(i)}, W_l^{(i)}\}$ and Weiss Green's functions $\{\mathcal{G}_\sigma^{(i)}, \mathcal{F}^{(i)}\}$. Using ED we calculate the self-energy of the system, which allows us to calculate new Weiss Green's function as it is described in section 3.3. Within ED we also calculate new parameters for the SIAM. A difference with the standard DMFT loop is that instead of calculating the Weiss Green's functions from the ED, we take the value of Weiss Green's functions calculated via a R-DMFT loop in the previous R-DMFT iteration.

3.4.2 Numerical Renormalization Group (NRG)

Another impurity solver for the single impurity Anderson model (SIAM) which we are using in this thesis is the *numerical renormalization group* (NRG). This method describes low-frequency behavior much better than exact diagonalization. One of the things which one has to take into account for solving the SIAM is that all energy scales contribute to the solution. To solve the problem one has to find non-perturbative approach. The first step in this direction was made by Anderson [186] with his *Poor Man's Scaling* approach. In this work he integrated out electron states close to the band edges of the conduction band, which have higher eigenenergies compared to the other states. He described the system of the remaining part of the conduction band with an effective Hamiltonian having the same mathematical form as initial one and calculated new coupling constant J' in a perturbation series considering only leading term. He showed that for the temperatures less than Kondo temperature T_K, the system is in the strong coupling regime, i.e. the impurity is strongly coupled to the conduction band electrons, but using his method it is not possible to investigate the system further. Soon after Anderson's work, in 1975 Wilson invented the numerical renormalization group (NRG) technique [170]. In this work the NRG was applied to the Kondo model, which one can get directly from the Anderson model using the Schrieffer-Wolff transformation [187], freezing charge (fermionic number) fluctuations. Later the NRG technique was extended also to the SIAM [172, 173].

Consider the SIAM Hamiltonian:

$$\hat{\mathcal{H}}_{\text{SIAM}} = \hat{\mathcal{H}}_{imp} + \sum_{l,\sigma} \left\{ \varepsilon_{l\sigma} \hat{a}_{l\sigma}^\dagger \hat{a}_{l\sigma} + V_{l\sigma} \left(\hat{f}_\sigma^\dagger \hat{a}_{l\sigma} + h.c. \right) \right\}. \qquad (3.46)$$

Here $\hat{\mathcal{H}}_{imp}$ describes the impurity which depending on the model can have different forms. l labels the noninteracting orbitals of the effective bath, $\varepsilon_{l\sigma}$ is the energy of the noninteracting orbital l for spin σ and $V_{l\sigma}$ is the corresponding fermionic hybridization matrix element. Here we assumed that energy of the orbital $\varepsilon_{l\sigma}$ does not depend on direction of **k** momentum, i.e. we consider only s-wave states.

We replace summation over l with an integral over energy:

$$\hat{\mathcal{H}}_{\text{SIAM}} = \hat{\mathcal{H}}_{imp} + \sum_\sigma \int_{-D}^{D} d\varepsilon \left\{ \varepsilon \hat{\alpha}_{\varepsilon\sigma}^\dagger \hat{\alpha}_{\varepsilon\sigma} + \sqrt{\frac{\Delta_\sigma(\epsilon)}{\pi}} \left(\hat{f}_\sigma^\dagger \hat{\alpha}_{\varepsilon\sigma} + h.c. \right) \right\}, \qquad (3.47)$$

where D is the non-interacting fermionic half-band width. We now define a hybridization function

$$\Delta_\sigma(\epsilon) = \pi \sum_l \delta(\varepsilon - \varepsilon_{l\sigma}) V_{l\sigma}^2. \qquad (3.48)$$

The next step is a logarithmical discretization of the conducting band [170]. We divide the conducting band into exponentially decreasing intervals (Fig. 3.5) and perform a Fourier expansion in each of this intervals. One can define a complete set of orthonormal functions:

$$\Psi_{np}^\pm = \begin{cases} \frac{1}{v_n} e^{\pm \frac{2\pi i p \varepsilon}{v_n}} & \Lambda^{-(n+1)} < \pm \varepsilon < \Lambda^{-n} \\ 0 & \text{otherwise} \end{cases}, \qquad (3.49)$$

where $v_n = \Lambda^{-n}(1 - 1/\Lambda)$ and Λ is the logarithmical discretization parameter. Using these orthonormal functions we can express $\hat{\alpha}_{\varepsilon\sigma}$ as follows:

$$\hat{\alpha}_{\varepsilon\sigma} = \sum_{np} \left[\hat{a}_{np\sigma} \Psi_{np}^+ + \hat{b}_{np\sigma} \Psi_{np}^- \right], \qquad (3.50)$$

where

$$\hat{a}_{np\sigma} = \int_{-D}^{D} d\varepsilon [\Psi_{np}^+(\varepsilon)]^\star \hat{\alpha}_{\varepsilon\sigma} \quad \text{and} \quad \hat{b}_{np\sigma} = \int_{-D}^{D} d\varepsilon [\Psi_{np}^-(\varepsilon)]^\star \hat{\alpha}_{\varepsilon\sigma}. \qquad (3.51)$$

One can easily check that the new operators $\hat{a}_{np\sigma}$ and $\hat{b}_{np\sigma}$ fullfill fermionic commutation relations.

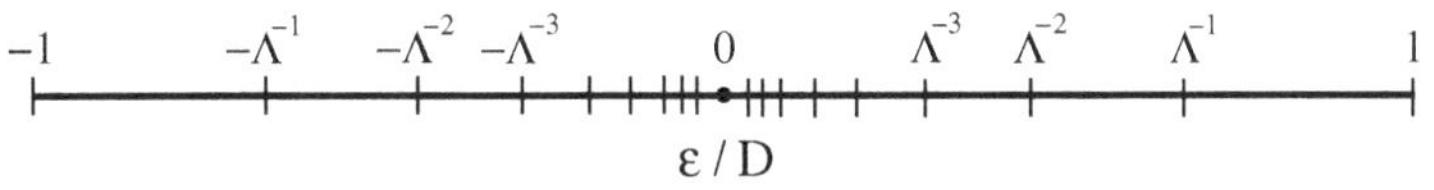

3.4 Impurity Solvers

Figure 3.5: Logarithmical discretization of the conduction band.

Using Eq. (3.50) one can rewrite the Hamiltonian (3.47) in new operators:

$$\hat{\mathcal{H}}_{SIAM} = \hat{\mathcal{H}}_{imp} + \sqrt{\frac{\zeta_{0\sigma}}{\pi}} \sum_\sigma (\hat{f}^\dagger_\sigma \hat{d}_{0\sigma} + h.c.) + \frac{1+\Lambda^{-1}}{2} \sum_{np\sigma} \Lambda^{-n}(\hat{a}^\dagger_{np\sigma}\hat{a}_{np\sigma} - \hat{b}^\dagger_{np\sigma}\hat{b}_{np\sigma})$$
$$+ \frac{1-\Lambda^{-1}}{2\pi i} \sum_{n\sigma, p \neq p'} \frac{\Lambda^{-n}}{p-p'} (\hat{a}^\dagger_{np\sigma}\hat{a}_{np'\sigma} - \hat{b}^\dagger_{np\sigma}\hat{b}_{np'\sigma}) e^{\frac{2\pi i(p-p')}{1-\Lambda^{-1}}}, \quad (3.52)$$

where

$$\hat{d}_{0\sigma} = \frac{1}{\zeta_{0\sigma}} \int_{-D}^{D} d\varepsilon \sqrt{\Delta_\sigma(\epsilon)} \hat{a}_{\varepsilon\sigma} = \frac{1}{\zeta_{0\sigma}} \sum_{n=0}^{\infty} (\gamma^+_{n\sigma}\hat{a}_{n0\sigma} + \gamma^-_{n\sigma}\hat{b}_{n0\sigma}) \quad (3.53)$$

and

$$\gamma^\pm_{n\sigma} = \pm \frac{1}{v_n} \int_{\pm\Lambda^{-(n+1)}}^{\pm\Lambda^{-n}} \sqrt{\Delta_\sigma(\epsilon)}, \qquad \zeta_{0\sigma} = \sum_n (\gamma^{+\,2}_{n\sigma} + \gamma^{-\,2}_{n\sigma}). \quad (3.54)$$

It is easy to verify that for $\Lambda \to 1$ the last term in the Hamiltonian (3.52) will disappear. This means that states with different quantum number p will be decoupled from each other and as the impurity is only coupled to the states with quantum number $p = 0$, we can neglect all states with $p \neq 0$. Wilson showed in his paper [170], that we can neglect these states even for a discretization parameter $\Lambda = 2$. So after neglecting the terms with $p \neq 0$ the Hamiltonian is given by the following equation:

$$\hat{\mathcal{H}}_{SIAM} = \hat{\mathcal{H}}_{imp} + \sqrt{\frac{\zeta_{0\sigma}}{\pi}} \sum_\sigma (\hat{f}^\dagger_\sigma \hat{d}_{0\sigma} + h.c.) + \frac{1+\Lambda^{-1}}{2} \sum_{\sigma, n=0}^{\infty} \Lambda^{-n}(\hat{a}^\dagger_{n\sigma}\hat{a}_{n\sigma} - \hat{b}^\dagger_{n\sigma}\hat{b}_{n\sigma}). \quad (3.55)$$

We observe that the energy scales are now separated and each state is described by two degrees of freedom $\hat{a}_{n\sigma}$ and $\hat{b}_{n\sigma}$. Wilson suggested to solve the problem by iterative perturbation theory: because the terms in the conduction band are now exponentially decreasing, they can be taken into account one after other [170]. In order to do this one has to perform an unitary transformation from the set of parameters $\{\hat{a}_{n\sigma}, \hat{b}_{n\sigma}\}$ to new operators $\{\hat{d}_{n\sigma}\}$. One can choose the new basis in such a way that the operator $\hat{d}_{n\sigma}$ is coupled to $\hat{d}_{n\pm1,\sigma}$. This can be achieved using a Lanczos tridiagonalization procedure [188]. The resulting

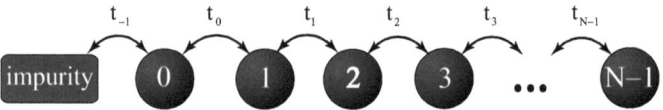

Figure 3.6: Schematic structure of the NRG chain. $t_{-1} = \sqrt{\frac{\zeta_{0\sigma}}{\pi}}$ describes hopping from the "impurity" site to the 0-th site of the chain.

Hamiltonian has the form of a semi-infinite linear chain (see Fig.3.6):

$$\hat{\mathcal{H}}_{SIAM} = \hat{\mathcal{H}}_{imp} + \sqrt{\frac{\zeta_{0\sigma}}{\pi}} \sum_\sigma (\hat{f}^\dagger_\sigma \hat{d}_{0\sigma} + h.c.) + \sum_{\sigma,n=0}^\infty t_{n\sigma}(\hat{d}^\dagger_{n\sigma}\hat{d}_{n+1,\sigma} + h.c.) + \sum_{\sigma,n=0}^\infty \delta_{n\sigma}\hat{d}^\dagger_{n\sigma}\hat{d}_{n\sigma}.$$
(3.56)

We would like to note that $t_{n\sigma} \sim \Lambda^{-n/2}$ and $\delta_{n\sigma} \sim \Lambda^{-n/2}$, which allows us to neglect terms far away from the impurity site in the linear chain. So we can consider the finite size Hamiltonian with N sites in the chain. Here we would like to note that this is the cut-off in the energy and is different from the finite-size problem of a finite number of lattice sites of the original model.

$$\hat{\mathcal{H}}_N = \Lambda^{(N-1)/2}\left[\hat{\mathcal{H}}_{imp} + \sqrt{\frac{\zeta_{0\sigma}}{\pi}} \sum_\sigma (\hat{f}^\dagger_\sigma \hat{d}_{0\sigma} + h.c.) + \sum_{\sigma,n=0}^{N-1} t_{n\sigma}(\hat{d}^\dagger_{n\sigma}\hat{d}_{n+1,\sigma} + h.c.) + \sum_{\sigma,n=0}^{N} \delta_{n\sigma}\hat{d}^\dagger_{n\sigma}\hat{d}_{n\sigma}\right].$$
(3.57)

In Eq. (3.57) we rescale the Hamiltonian $\hat{\mathcal{H}}_N$ such that low-lying excitations are always of order one. The Hamiltonian of the full system is recovered if one considers the following limit:

$$\hat{\mathcal{H}}_{SIAM} = \lim_{N\to\infty} \Lambda^{-(N-1)/2}\hat{\mathcal{H}}_N.$$
(3.58)

To solve the problem one can use the *iterative diagonalization*. It is based on the fact that:

$$\hat{\mathcal{H}}_{N+1} = \Lambda^{1/2}\hat{\mathcal{H}}_N + \Lambda^{N/2}\sum_\sigma t_{N\sigma}(\hat{d}^\dagger_{N\sigma}\hat{d}_{N+1,\sigma} + h.c) + \Lambda^{N/2}\sum_\sigma \delta_{N+1,\sigma}\hat{d}^\dagger_{N+1,\sigma}\hat{d}_{N+1,\sigma}.$$
(3.59)

Once we know the eigenvalues and eigenvectors of the Hamiltonian $\hat{\mathcal{H}}_N$, we can also calcu-

3.4 Impurity Solvers

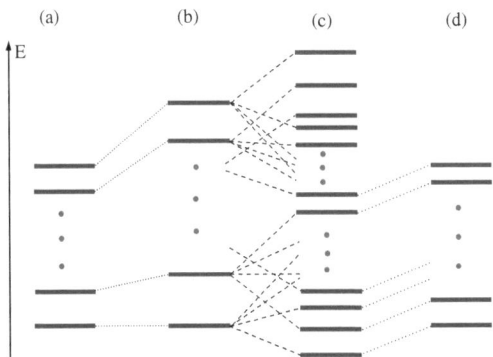

Figure 3.7: (a) The N_{level} lowest levels of the Hamiltonian $\hat{\mathcal{H}}_N$ after N iterations. The ground state energy is set to zero. (b) The same levels after rescaling: $E_l \to \sqrt{\Lambda}E_l$. (c) The levels of the Hamiltonian $\hat{\mathcal{H}}_{N+1}$ after $N+1$ iteration, each level split into 4 different levels. (d) The same levels as in (c) but after truncation. We are keeping the N_{level} lowest levels. The ground state energy is again set to zero.

late matrix elements of $\langle \hat{d}_{N\sigma} \rangle_{ij}$ and build the new Hamiltonian for site $N+1$ (for details see Appendix F). Adding one more site the size of the Hamiltonian matrix is increasing by a factor of 4 for spin-1/2 fermions and by a factor of 2 for spinless fermions (see Fig. 3.7). To avoid exponential increase of the matrix size, we are keeping not all levels but only the N_{level} lowest levels (see Fig. 3.7). To speed up the diagonalization process we are using the symmetries of the model, which allows us to write the Hamiltonian in a block diagonal structure. Depending on the problem, the total number of fermions N_f or/and the total magnetization M_z is conserved. When both of these quantum numbers are conserved, the size of each block in the Hamiltonian matrix is much smaller, compared to the case when only one of the quantum numbers (N_f, M_z) is conserved.

An impurity solver for (G)DMFT calculation has to be able to calculate correlation functions and Green's functions for the "impurity site". For this purpose we again use the iterative diagonalization. We first calculate matrix elements of the desired operators on the "impurity site", then adding one more site we transform these quantities and obtain new values from them. We continue this process until we consider all N_{site} sites of the linear chain (for details see Appendix F).

Let us now briefly explain how NRG works as an impurity solver for the DMFT algorithm. We start from an initial guess of the hybridization function, then knowing the hybridization function we can calculate hopping coefficients for the NRG chain. Afterwards we perform the NRG calculation, which allows us to calculate spectral and correlation functions. Working on a Bethe lattice we can then directly calculate the hybridization function:

$$\Delta_\sigma(\omega) = \pi t_f^{*2} A_\sigma(\omega), \qquad (3.60)$$

where t_f^* is the rescaled hopping coefficient of the original lattice, and the spectral function is given by

$$A_\sigma(\omega) = -\frac{1}{\pi}\Im m G_\sigma(\omega) = \frac{1}{Z}\sum_{nm}|\langle m|\hat{f}_\sigma^\dagger|n\rangle|^2 \delta(\omega - E_m + E_n)\left(e^{-\beta E_n} + e^{-\beta E_m}\right), \qquad (3.61)$$

where Z is the statistical sum, $|n\rangle$ is a many particle eigenstate of the $\hat{\mathcal{H}}_N$ Hamiltonian and E_n is the corresponding energy eigenvalue.

For more complicated lattices, in addition to the spectral function A_σ, we also calculate the interacting spectral functions

$$B_{ff,\sigma}(\omega) = -\frac{1}{\pi}\Im m Q_{ff,\sigma}(\omega) = \frac{1}{Z}\sum_{nm}\langle n|\hat{f}_\sigma \hat{f}_{\bar\sigma}^\dagger \hat{f}_{\bar\sigma}|m\rangle\langle m|\hat{f}_\sigma^\dagger|n\rangle \delta(\omega - E_m + E_n)\left(e^{-\beta E_n} + e^{-\beta E_m}\right) \qquad (3.62)$$

and for Bose-Fermi mixtures in addition we have to calculate

$$B_{fb,\sigma}(\omega) = -\frac{1}{\pi}\Im m Q_{fb,\sigma}(\omega) = \frac{1}{Z}\sum_{nm}\langle n|\hat{f}_\sigma \hat{b}^\dagger \hat{b}\,|m\rangle\langle m|\hat{f}_\sigma^\dagger|n\rangle \delta(\omega - E_m + E_n)\left(e^{-\beta E_n} + e^{-\beta E_m}\right), \qquad (3.63)$$

where $\bar\sigma = -\sigma$. After calculating the spectral functions A_σ, $B_{ff,\sigma}$ and $B_{fb,\sigma}$, using the Kramers-Kronig relation one can calculate the Green's functions $G_\sigma(\omega)$, $Q_{ff,\sigma}(\omega)$ and $Q_{fb,\sigma}(\omega)$. Knowing these Green's functions we can then calculate the self-energy:

$$\Sigma_\sigma = U_f \frac{Q_{ff,\sigma}}{G_\sigma} + U_{fb}\frac{Q_{fb,\sigma}}{G_\sigma} \qquad (3.64)$$

and the DMFT loop is closed by the Dyson equation (3.10). The imaginary part of the Weiss Green's function is the hybridization function.

Chapter 4

Mixtures of Fermions and Bosons in Optical Lattices

Ultracold atomic gases allow to realize novel quantum many-body systems. In particular, one can perform experiments on Bose-Fermi mixtures [4–19]. One of the key questions that has been explored is the effect of fermions on the mobility of the bosons. When fermions are slow compared to the bosons, they act as dynamical impurities. Fast fermions, on the other hand, mediate long-range interactions between the bosons. In both cases this

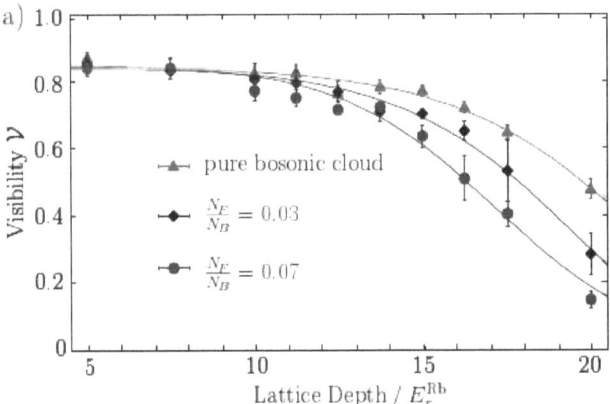

Figure 4.1: Visibility of the bosonic interference pattern for different fermionic impurity concentrations for a mixture of ^{40}K and ^{87}Rb. (From Ref. [12]).

has an effect on the superfluid-Mott insulator transition. Different experimental groups intensively studied this shift of the superfluid-Mott insulator transition induced by the fermions [10–16]. The coherence properties of the bosonic cloud can be revealed by studying the bosonic interference pattern after an instantaneously switching off the lattice in a time-of flight experiment. For this purpose the visibility of the interference fringes has to be investigated [16, 189, 190]. Experiments show that adding fermionic atoms decreases the visibility (See Fig. 4.1), which indicates that adding fermions leads to a stabilization of the Mott insulator phase. There are several theoretical studies which try to explain this phenomenon [22, 32–34, 42]. However, so far they predict that within a single band approximation, the Mott region shrinks, which is at variance with experimental results. Multiband-effects however, can lead to an extension of the Mott insulating regime [43–45].

In this chapter we will consider mixtures of fermions and bosons in optical lattices for commensurate filling of both fermions and bosons at zero temperature. Our results will serve as predictions for future experiments. In our calculations we are neglecting the effect of the harmonic trap.

In particular, first we will consider a mixture of spinless fermions and bosons in two different cases: (i) when the filling of fermions and bosons is 1/2 (section 4.1.1) and (ii) when the spinless fermions are half-filled while the filling of the bosons is 3/2 (section 4.1.2). In section 4.2 we will consider the case of a mixture of bosons and two-component fermions when both fermions as well as bosons are half-filled.

We remark here that the results presented in the current chapter are obtained with a density of states without Van Hove singularities. In fact, the results were obtained using the density of states of the Bethe lattice, which is semi-elliptic and regular everywhere (see Eq. 3.11). As we will show below we were able to identify a *supersolid phase* - the phase with coexisting broken $U(1)$ symmetry and particle density wave order - proving the point that a singularity in the non-interacting states is not a necessary condition for the occurrence of a supersolid.

During our calculations we use the *GDMFT*. As an impurity solver we use the *Numerical renormalization group* (NRG). We use the following NRG parameters: logarithmical discretization $\Lambda = 2$, number of the NRG iterations $N_{iter} = 60$, the number of the kept states $N_{level} = 1000$ and the bosonic cut-off 4 for half-filled bosons (section 4.1.1) and the bosonic cut-off 6 for 3/2-filled bosons (section 4.1.2). Obviously the bosonic cut-off for hard-core bosons (section 4.2) will be 1, because in this limit each site can be occupied by at most one boson.

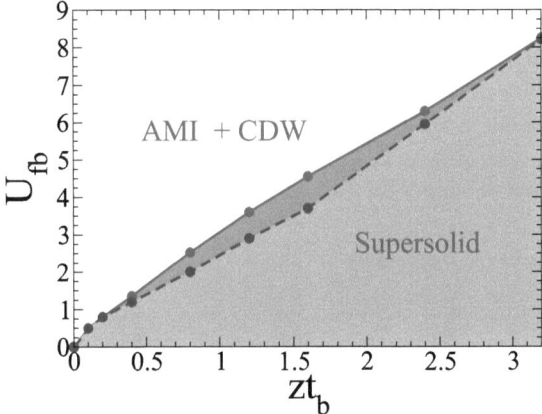

Figure 4.2: Phase diagram of the Fermi-Bose Hubbard model with spinless fermions and hard-core bosons at half filling. We identify the supersolid phase (below the red solid line), the alternating Mott insulator (AMI) phase with charge density wave (CDW) (above red line), and the coexistence region (between the red and dashed blue line). Energies are expressed in units of the non-interacting fermionic half-bandwidth D.

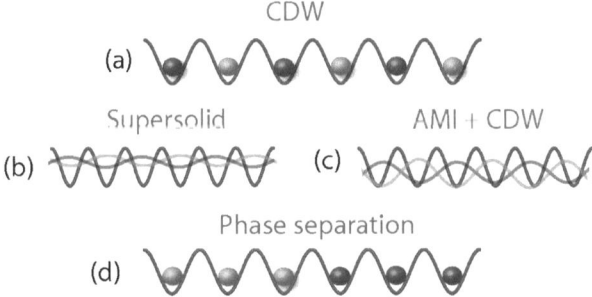

Figure 4.3: Schematic structure of different $T = 0$ phases of a spinless Bose-Fermi mixture in an optical lattice. The red (dark) particles correspond to fermions, while cyan (light) particles denote bosons. In both the supersolid and the AMI phase the bosons and fermions have an alternating density pattern as depicted in (a). In the supersolid (b) the density oscillations are small and the bosons are superfluid. In the AMI + CDW phase (c) the density oscillations are large and the bosons are localized. The schematic structure of phase separation is depicted in (d).

4.1 Mixtures of Spinless Fermions and Bosons in Optical Lattices

In this section we consider a mixture of spinless fermions and bosons in an optical lattice. Earlier theoretical studies already suggested that Bose-Fermi mixtures can be unstable against charge density wave (CDW) and supersolid order or phase separation (PS). However, so far all theoretical approaches either dealt with one-dimensional systems [22–28], or relied on weak-coupling approximations [29–31].

4.1.1 Half-Filled Mixture of Spinless Fermions and Bosons

A mixture of spinless fermions and bosons in an optical lattice can be described by the single-band Fermi-Bose Hubbard model:

$$\hat{\mathcal{H}} = -\sum_{\langle i,j \rangle} \left\{ t_f \hat{c}_i^\dagger \hat{c}_j + t_b \hat{b}_i^\dagger \hat{b}_j \right\} - \sum_i \left\{ \mu_f \hat{n}_i^f + \mu_b \hat{n}_i^b \right\} + \sum_i \left\{ \frac{U_b}{2} \hat{n}_i^b (\hat{n}_i^b - 1) + U_{fb} \hat{n}_i^b \hat{n}_i^f \right\}, \quad (4.1)$$

where $\hat{c}_i^\dagger$ ($\hat{b}_i^\dagger$) is the fermionic (bosonic) creation operator at site i, while $\hat{n}_i^f = \hat{c}_i^\dagger \hat{c}_i$ ($\hat{n}_i^b = \hat{b}_i^\dagger \hat{b}_i$) denotes the number operator and $\mu_{f(b)}$ the chemical potential for fermions (bosons). U_b and U_{fb} are the on-site boson-boson and fermion-boson interactions respectively. $\langle i,j \rangle$ denotes summation over nearest neighbors, and $t_{f(b)}$ is the tunneling amplitude for fermions (bosons).

We now first study the limit $U_b = \infty$, i.e. hard-core bosons. In this limit each site can be occupied only by a single boson. In one dimension, if the hard-core bosons hop only to nearest neighbor sites, one can make a unitary transformation, and express hard-core bosons in terms of non-interacting fermions. So, the behavior of hard-core bosons is similar to the behavior of fermions. However, in higher dimensions one cannot make such a unitary transformation to map hard-core bosons to non-interacting fermions. Correspondingly, the behavior of hard-core bosons and fermions is then not the same.

As we mentioned above, we consider the case when both bosons and spinless fermions are half-filled ($\langle \hat{n}_b \rangle = \langle \hat{n}_f \rangle = \frac{1}{2}$), which makes the system particle-hole symmetric. This is ensured by the choice of the chemical potentials equal to $\mu_f = \mu_b = U_{fb}/2$. Without loss of generality, calculations are performed for repulsive Fermi-Bose interactions: $U_{fb} > 0$. The case of attractive interactions will be inferred later on with the help of a (staggered) particle-hole transformation.

4.1 Mixtures of Spinless Fermions and Bosons in Optical Lattices

We take the non-interacting fermionic half-bandwidth $D = 2t_f^*$ as the unit of energy, the bosonic hopping amplitude t_b and the interaction U_{fb} are the remaining adjustable parameters. Our results are shown in the $U_{fb} - t_b$ phase diagram in Fig. 4.2. In Fig. 4.3, we schematically depict each of these phases.

For weak repulsion between fermions and bosons we obtain a *supersolid phase* (the blue area below the blue dashed line in Fig. 4.2, schematically we depict this phase in Fig. 4.3(a,b)). In the supersolid phase both the fermions and the bosons form a Charge Density Wave (CDW) and the bosons are superfluid, i.e. this is the phase with coexisting broken $U(1)$ symmetry and particle density wave order. In the Fig. 4.4 we plot the amplitude of the CDW as a function of U_{fb}. As one can see the amplitude of the CDW oscillation is small in the supersolid phase. For strong interactions between fermions and bosons we obtain a bosonic *alternating Mott insulator phase* (AMI) together with a charge density wave (CDW) of the fermions (upper part of the phase diagram above the red solid line in Fig. 4.2, schematically shown in Fig. 4.3(a,c)). In this phase the fermionic CDW amplitude $|\Delta N_f| \equiv |N_f - \langle \hat{n}_f \rangle|$ is almost maximal, while the bosons are completely localized and have a CDW amplitude equal to $|\Delta N_b| \equiv |N_b - \langle \hat{n}_b \rangle| = 0.5$. Taking into account virtual bosonic particle-hole excitations beyond the Gutzwiller approximation [161, 162] would, however, lead to a slightly smaller bosonic CDW amplitude. This transition is very similar to the one for bosons in a superlattice: upon increasing the potential difference between the sublattices there is a Mott-insulator transition at half filling [191–193]. For intermediate coupling (in the area between the red solid line and blue dashed line in Fig. 4.2) both solutions are stable within GDMFT. To determine which of them corresponds to the ground state, we have compared their energies as given by Eq. (3.33). We find that the supersolid phase always has the lower energy, i.e. in Fig. 4.2 between the solid red line and the dashed blue line the ground state is the supersolid. The coexistence of GDMFT solutions is a strong indication for a first order phase transition (at $T = 0$). Another strong indication for a first order phase transition is a jump in the CDW amplitude for fermions and bosons (see Fig. 4.4). As shown in Fig. 4.2, the critical value U_{fb}^c for the phase transition from the supersolid into the AMI phase increases with the bosonic tunneling amplitude.

We also study the fermionic spectrum and find that the fermionic spectrum is always gapped. Spectral densities are shown in Fig. 4.5. The gap is small for the supersolid phase, but at the transition point there is a jump in the gap and in the AMI phase it becomes of the order of the non-interacting half-bandwidth D (see inset of Fig. 4.5). This implies that the latter phase will be more stable against finite temperature effects.

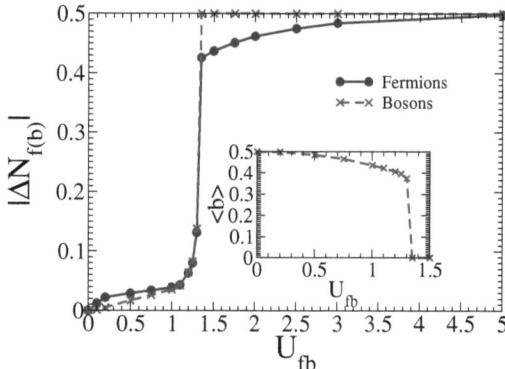

Figure 4.4: Amplitude of the CDW for fermions (blue circles, solid line) and hard-core bosons (red crosses, dashed line) as a function of the fermion-boson interaction U_{fb} for the case when $zt_b = 0.4D$. In the inset we plot the bosonic superfluid order parameter as a function of the fermion-boson interaction U_{fb}.

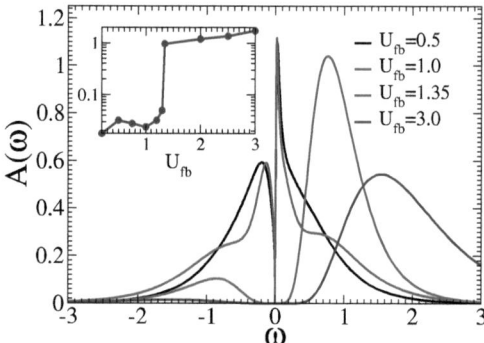

Figure 4.5: Fermionic spectral function in a mixture with hard-core bosons and $zt_b = 0.4D$ for different values of U_{fb}. In the inset we plot the size of the gap in units of D as a function of U_{fb}. The gap is defined by the frequencies for which the spectral function has half its maximal value.

4.1 Mixtures of Spinless Fermions and Bosons in Optical Lattices   61

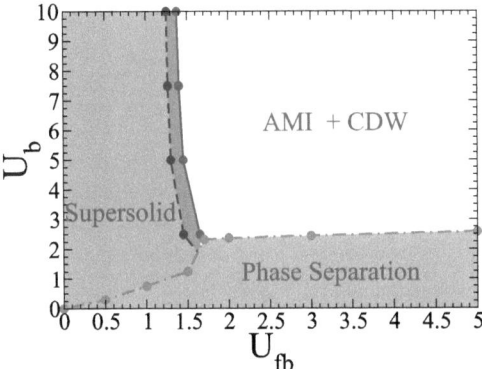

Figure 4.6: Phase diagram of the Fermi-Bose Hubbard model with spinless fermions. Both fermions and bosons are half-filled and $zt_b = 0.4D$. Stable phases are the supersolid (left of the red solid line and above the green dash-dotted line) and the alternating Mott insulator (AMI) phase with charge density wave (CDW) (right of the red solid line and above the green dash-dotted line). In the area between the red solid line and blue dashed line both solutions are stable. Below the green dash-dotted line phase separation (PS) takes place.

So far we have considered repulsive interactions between bosons and fermions. To see what happens for attractive interactions $U_{fb} < 0$ we apply a staggered particle-hole transformation to the fermions, $\hat{c}_i \to (-1)^i \hat{c}_i^\dagger$, which leads to a minus sign in front of the Bose-Fermi interaction term. This implies that for attractive interactions we obtain the same quantum phases, but the CDW-oscillations are now in-phase, instead of out-of-phase as for repulsive interactions.

We now proceed by considering finite interactions between the bosons, i.e. relaxing the hard-core condition, but still assume the fermions and the bosons to be half-filled. In this case we have to adjust chemical potential for both fermions and bosons to get the correct filling. We consider the case that the bosons are slightly slower than the fermions: $zt_b = 0.4D$. Our findings are summarized in the $U_{fb} - U_b$ phase diagram in Fig. 4.6. For strong bosonic repulsion U_b the results are similar to the ones found for hard-core bosons: we find a supersolid for weak U_{fb} and the alternating Mott insulator phase for stronger U_{fb}, separated by a first order transition. Also in this case we find a region where both GDMFT solutions are stable, but the supersolid state is lower in energy than the AMI. The critical interspecies repulsion at the transition between supersolid and the AMI phase increases when the value of the bosonic repulsion U_b is reduced. This is because the supersolid state acquires a lower energy when U_b is decreased, whereas the energy of the AMI phase remains the same. For weak interactions U_b among the bosons, the half-filled state is unstable towards *Phase Separation* (PS) (green area below green dash-dotted line in Fig. 4.2, schematically we depict this phase in Fig. 4.3(d)). In this parameter regime we do not find a converged GDMFT solution where the bosons and the fermions are half-filled. To establish the occurrence of phase separation we also performed calculations away from half filling. We found a pronounced jump in the density as a function of the chemical potential and coexisting solutions close to the position of the jump. Moreover, we observed that for strong interspecies repulsion the phase separation is always complete. This allowed us to compare the energies of the PS- and AMI states, which yields the green dash-dotted line as depicted in Fig. 4.6. We have checked that comparison of energies yields the same boundary for phase separation as deduced from the disappearance of a converged homogeneous GDMFT solution.

Also in this case we can infer the effect of attractive Bose-Fermi interactions by performing a staggered particle-hole transformation for the fermions. Phase separation turns then into phase separation of bosons and fermionic holes, which is equivalent to clustering of the bosonic and fermionic particles. So for weak repulsion U_b among the bosons a system with attractive interspecies interaction U_{fb} will maximize its density in part of the system, leaving the rest unoccupied.

4.1.2 3/2-Filled Bosons and Half-Filled Spinless Fermions

In the previous subsection, we established the existence of a supersolid phase at filling $N_b = N_f = 1/2$. However, the particle density oscillation and the gap in the spectrum in the supersolid phase were rather small. This is partly due to the fact that the Bethe lattice density of state is regular everywhere, and has not singularities which could enhance the effects. However, this makes the experimental verification of this phase very challenging.

Therefore, in this section we study a different case where the filling of spinless fermions is 1/2, while the filling of the bosons is higher, namely $\langle \hat{n}_i^b \rangle = 3/2$. The reason for this particular choice is that it allows for two different Alternating Mott Insulator (AMI) phases, with amplitude of the bosonic density oscillation 1/2 and 3/2, respectively. These two AMI phases are separated by a supersolid phase. The amplitude of the density oscillations in this supersolid phase in between the two AMI phases is of the order one, which makes its experimental detection much easier.

To overcome the tendency towards phase separation in the system, we consider the case

4.1 Mixtures of Spinless Fermions and Bosons in Optical Lattices   63

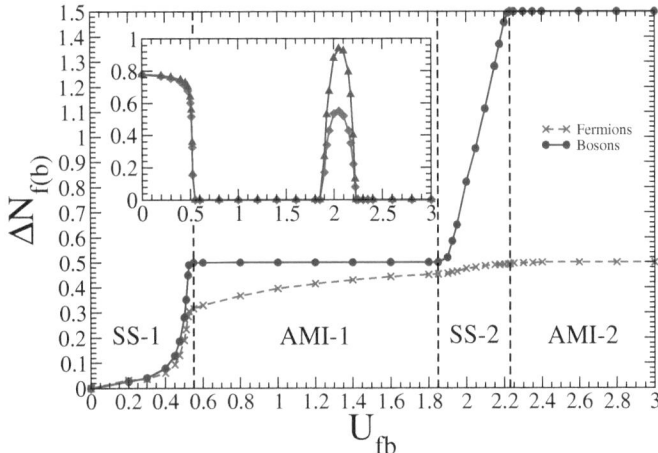

Figure 4.7: Dependence of the amplitude of the bosonic/fermionic density wave on the Fermi-Bose interaction, for the case when $zt_b = 0.05D$ and $U_b = 1.0D$, where D denotes the half-band width of the fermions. In the inset we depict the superfluid order parameter. The different line types in the inset correspond to results on the two sublattices. The different phases are schematically depicted in Fig. 4.8.

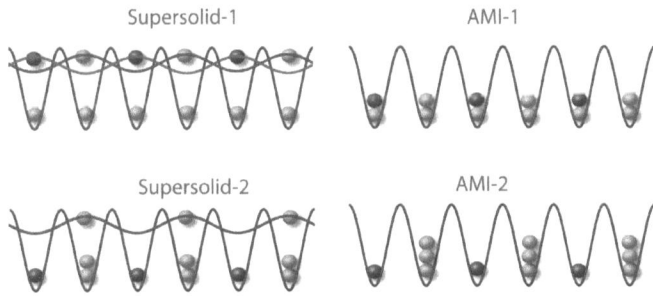

Figure 4.8: Schematic picture of the four different phases occurring in the Bose-Fermi mixture for bosonic filling 3/2 and fermionic filling 1/2. We identify the Supersolid-1 phase in which superfluidity coexists with a charge density wave with $\Delta N_b < \frac{1}{2}$. The AMI-1 has localized bosons with $\Delta N_b = \frac{1}{2}$. The Supersolid-2 phase is defined by superfluidity coexisting with a charge density wave with $\frac{1}{2} < \Delta N_b < \frac{3}{2}$. The AMI-2 has localized bosons with $\Delta N_b = \frac{3}{2}$.

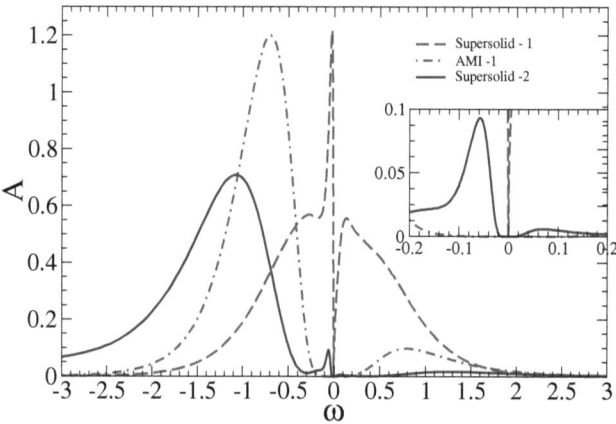

Figure 4.9: The fermionic spectral functions for the different phases. The parameters are chosen the same as in Fig. 4.7. The dashed green line corresponds to the supersolid-1 phase ($U_{fb} = 0.4D$), the dash-dotted red line corresponds to the AMI-1 phase with bosonic CDW oscillation 0.5 ($U_{fb} = D$) and the blue line corresponds to the supersolid-2 phase ($U_{fb} = 1.95D$). In the inset we plot the same spectral functions, at smaller frequencies.

where the bosons are much slower than the fermions $zt_b = 0.05D$, and where the repulsion among the bosons is strong $U_b = D$.

GDMFT analysis

First we study this situation by means of GDMFT. In Fig. 4.7, we plot the amplitude of the density oscillations as a function of the interspecies interaction U_{fb}. The amplitude of the density oscillations is defined as $\Delta N_{f(b)}$.

The results show that the oscillation amplitude is a smooth function of U_{fb} for fermions and bosons. We identify four different regimes in the system. Schematic pictures for these four phases are given in Fig. 4.8. For weak interactions between fermions and bosons the system is in the supersolid phase: the bosons are superfluid and there is a spontaneous particle density oscillation in the system, which increases with increasing interaction U_{fb}. For some critical U_{fb} the bosonic density amplitude reaches 1/2. At this point, the system undergoes a transition into the AMI-1 phase. Here the bosonic density is alternating between 1 and 2 on neighboring lattice sites. If we continue to increase the interaction, only the amplitude of the fermionic density oscillations slowly increases. This continues

4.1 Mixtures of Spinless Fermions and Bosons in Optical Lattices

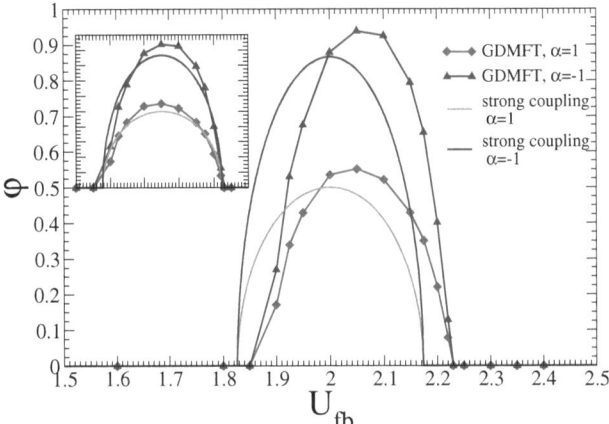

Figure 4.10: Superfluid order parameter on the two sublattices ($\alpha = \pm 1$) as a function of the Fermi-Bose interaction, obtained by means of the GDMFT and the strong coupling model. Parameters are chosen the same as in Fig. 4.7. In the inset we plot the same data, but the strong coupling results are shifted towards stronger U_{fb} to compensate for the screening caused by virtual hopping processes of the fermions, which are not included in the strong coupling model.

up to the second phase transition from the AMI phase into second supersolid phase. In this region, with increasing U_{fb}, both amplitudes of the density oscillations of fermions and bosons continuously increase, until the amplitude of the bosonic density oscillations reaches 3/2. At this point a phase transition occurs from the supersolid into a second AMI phase. Within this AMI-2 phase the bosons order themselves by alternating 0 and 3 bosons per site. Upon further increase of the interspecies interaction, the bosonic density oscillation - within our approximation - does not change, while the amplitude of the fermionic density oscillations converges to 1/2. In contrast to the case of half-filled hard-core bosons, the superfluid order parameter is different on the two sublattices for this case, because there is no particle-hole symmetry for the bosons. This is visible in the inset of Fig. 4.7, where the superfluid order parameter on the two sublattices is plotted.

An important observation concerns the order of the phase transitions. In the case of half-filled bosons, as we showed in the previous section the transition between the supersolid and the AMI phase is a first order quantum phase transition. However, for the bosonic density of 3/2, we find the transition to be of second order, as can be inferred from the lack of coexisting phases and the smooth behavior of all order parameters.

We also study the local spectral functions in the different phases. The results are displayed in Fig. 4.9. The gap in the first supersolid phase is very small, as also found for the supersolid phase with half-filled bosons. In the AMI phases we find that the fermions have a rather large gap. A more interesting structure emerges in the spectral function of the second supersolid phase. In this phase, in addition to the Hubbard sub-bands, an additional peak arises in the spectral function. The gap in this phase, as one can see from the inset of Fig. 4.9, is much larger than the gap in the first supersolid phase, but is also much smaller than in AMI phase. We have investigated the nature of the excitations responsible for this additional peak. These excitations correspond to a breaking of the alternating boson-fermion order in the system and therefore indicate the instability of the system to phase separation, which has only a slightly higher energy. In the AMI phase this energy difference is higher than in the supersolid phase, because the superfluid order parameter in the supersolid is oscillating (as seen from the inset of Fig. 4.7) and therefore reduced. This leads to an increase of the energy and therefore enhances the instability towards phase separation.

Strong coupling

To gain a better analytic understanding of the system, we also consider a strong coupling approach. We propose a simple model, where in one of the sublattices on each site a fermion is localized, whereas the sites of the other sublattice are occupied by localized pairs of bosons. In addition, we consider half-filled bosons on top of this arrangement. Within this model the AMI-1 phase is described by the localization of the additional bosons on the "fermionic" sublattice. The AMI-2 phase corresponds to localization in the sublattice with the boson-pairs. The supersolid corresponds to the case where the additional bosons are superfluid and delocalized over all lattice sites. To describe the phase transition within this strong coupling model, we have to study localization of half-filled bosons in a superlattice. The effective Hamiltonian in the Gutzwiller approach describing this situation has the form $\hat{\mathcal{H}}_{eff} = \frac{L}{2}\left(\hat{\mathcal{H}}_{-1} + \hat{\mathcal{H}}_{1}\right)$, where L is the number of lattices sites and

$$\hat{\mathcal{H}}_{1} = -zt_b\varphi_{-1}\left(\hat{a}_1^\dagger + \hat{a}_1\right) - (U_b - \frac{U_{fb}}{2})\left(\hat{n}_1 - \frac{1}{2}\right), \tag{4.2}$$

$$\hat{\mathcal{H}}_{-1} = -zt_b\varphi_1\sqrt{3}(\hat{a}_{-1}^\dagger + \hat{a}_{-1}) + (U_b - \frac{U_{fb}}{2})(\hat{n}_{-1} - \frac{1}{2}), \tag{4.3}$$

where the index ± 1 corresponds to the two sublatticies. The sublattice marked by 1 is occupied by localized fermions and on each site of sublattice -1 there are two localized bosons. We have treated the additional boson as hard-core, which is justified because of the large bosonic interaction U_b. The factor $\sqrt{3}$ comes from the fact that in the sublattice -1 we have three bosons. We solve this system self-consistently and find the values when this system has a non-trivial solution ($\varphi_{\pm 1} \neq 0$). Our result shows that the system is superfluid in the following range:

$$2U_b - 2\sqrt{3}zt_b < U_{fb} < 2U_b + 2\sqrt{3}zt_b.$$

Also we compare the superfluid order parameter calculated by strong coupling and the GDMFT (see Fig. 4.10). Our results show good agreement between these two results. Compared to the GDMFT-results, the strong coupling data are shifted towards smaller Bose-Fermi interaction. This shift is due to screening caused by the fact that in the superfluid phase the fermions are completely localized at the one sublattice, as we assumed in this strong-coupling argument. In reality, due to virtual hopping processes, there is also a finite density of fermions on the other sublattice. This effectively reduces the interaction between fermions and bosons.

4.2 Mixtures of Hard-Core Bosons and Two-Component Fermions in an Optical Lattice

In this section we will consider a mixture of bosons and two-component (i.e. spinful) fermions when filling of the fermions as well as the bosons is $1/2$. In our calculation we assume that the number of different components of the fermions are the same and that the bosons are hard-core.

Such a system can be described by the single-band Fermi-Bose Hubbard model:

$$\hat{\mathcal{H}} = -\sum_{\langle i,j \rangle, \sigma} \left\{ t_f \hat{c}_{i\sigma}^\dagger \hat{c}_{j\sigma} + t_b \hat{b}_i^\dagger \hat{b}_j \right\} - \sum_i \left\{ \mu_f \hat{n}_i^f + \mu_b \hat{n}_i^b \right\} + \sum_i \left\{ U_f \hat{n}_{i\uparrow}^f \hat{n}_{i\downarrow}^f + U_{fb} \hat{n}_i^b \hat{n}_i^f \right\}. \quad (4.4)$$

As before $\langle i,j \rangle$ denotes summation over nearest neighbors, $t_{f(b)}$ is the tunneling amplitude for fermions (bosons), $\hat{c}_i^\dagger$ ($\hat{b}_i^\dagger$) is the fermionic (bosonic) creation operator at the site i,

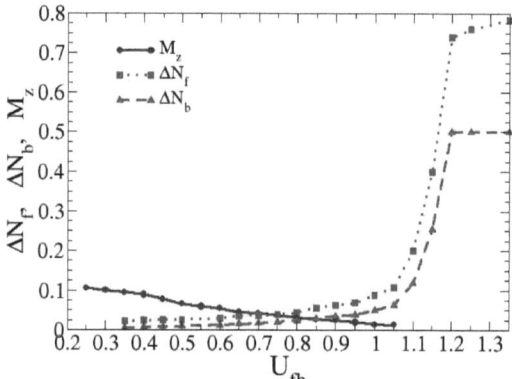

Figure 4.11: The staggered magnetization (blue circles, solid line), as well as amplitude of the CDW for fermions (red squares, dotted line) and hard-core bosons (green triangles, dashed line) as a function of the fermion-boson interaction. Here we would like to emphasize that magnetization and CDW amplitudes are corresponding to two different solutions.

while $\hat{n}^f_{i\sigma} = \hat{c}^\dagger_{i\sigma}\hat{c}_{i\sigma}$ ($\hat{n}^b_i = \hat{b}^\dagger_i\hat{b}_i$) denotes the number operator for fermions (bosons) and $\hat{n}^f_i = \hat{n}^f_{i\uparrow} + \hat{n}^f_{i\downarrow}$. $\mu_{f(b)}$ is the chemical potential for fermions (bosons). U_f and U_{fb} are the on-site fermion-fermion and fermion-boson interaction, respectively. Since we are considering hard-core bosons, two bosons can not occupy the same site and correspondingly do not interact.

First we would like to discuss the limiting case $U_{fb} = 0$. In this limit, fermions and bosons are decoupled from each other and the solution for each subsystem is well known. As mentioned, both fermions and bosons are half-filled and in this case the ground state for fermions and bosons corresponds to the *antiferromagnetic* and the *superfluid* phases, respectively.

For $U_f = 0$ the system is similar to the spinless fermion system which was described in section 4.1 and one can expect to obtain the *supersolid* and *alternating Mott insulator (AMI)* phases.

Here we would like to study the system for finite Fermi-Fermi and Bose-Fermi interactions. We again will take the non-interacting fermionic half-bandwidth $D = 2t^*$ as the unit of energy. In our calculations the bosons are slightly faster than the fermions: $zt_b = 0.6D$.

4.2 Mixtures of Hard-Core Bosons and Two-Component Fermions in an Optical Lattice

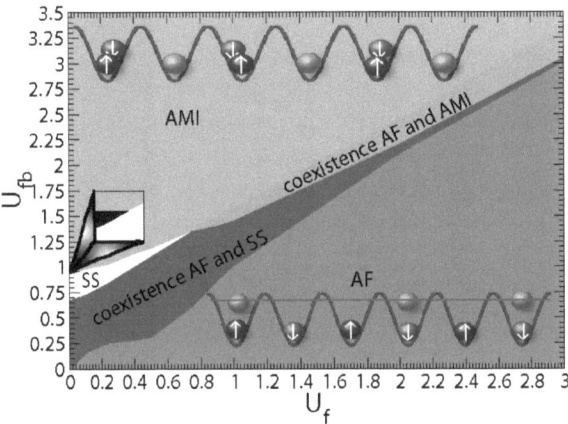

Figure 4.12: Phase diagram of the Fermi-Bose Hubbard model with two component fermions and hard-core bosons at half filling. We identify the antiferromagnetic phase (AF) (green area, in the bottom of the phase diagram), the alternating Mott insulator (AMI) phase with charge density wave (CDW) (cyan area in the top of the phase diagram, the supersolid phase (SS)(white area) and three coexistence regions between (i) the AF and SS phase (red area), (ii) the AF and AMI phase (orange area) and (iii) the SS and AMI (small black area).

Let us first consider the case when $U_f = 0.5D$. Our calculation shows that, when $U_{fb} < 0.35D$, we obtain the antiferromagnetic phase. We would like to point out that starting from any initial set of parameters we always end up with the antiferromagnetic solution. With increasing interaction ($U_{fb} \geq 0.35D$), we find either the antiferromagnetic or the supersolid phase depending on the initial conditions (see Fig. 4.11). With further increase of interaction ($U > 1.05D$) the antiferromagnetic solution disappears and we only get the supersolid solution. After further increase of interaction, a phase transition to the AMI phase takes place ($U_{fb} \simeq 1.2D$).

Our findings are summarized in the $U_{fb} - U_f$ phase diagram in Fig. 4.12. In the case when the interaction between fermions is stronger than the interaction between fermions and bosons, we obtain the antiferromagnetic phase, while in the opposite case, when interaction between fermions and bosons are stronger than interaction between fermions, we obtain the AMI phase. For the intermediate regime we obtain the supersolid phase and three different coexistence regions between (i) the AF and SS phase, (ii) the AF and AMI phase and (iii) the SS and AMI (see Fig. 4.12). As we have already discussed in the chapter 3, to find out which phase corresponds to the ground state one has to compare the energies of the coexisting phases. Unfortunately, the accuracy of our calculation at this point does not allow us to determine the state with the minimal energy.

Our key finding in this section is observation of the antiferromagnetic phase in the presence of the bosons. As the bosons can be used for cooling the system, this could facilitate the creation of an antiferromagnetic phase compared to a pure fermionic mixture.

Chapter 5

Ultra-Cold Atoms in a Harmonic Trap

The results presented in the previous chapter were obtained for a homogeneous system in the presence of the optical lattice. However, in experiments, the spatial inhomogeneity due to the harmonic confinement potential is always present, leading to a spatially varying local density. In this chapter we will take this effect into account. We again investigate the system at zero temperature. First we will consider a mixture of two-component fermions, while later on we will discuss a mixture of bosons and spinless fermions. In Fig. 5.1 we plot a schematic arrangment of ultra-cold atoms in a harmonic trap.

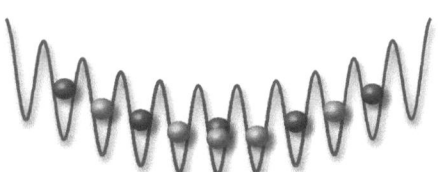

Figure 5.1: Schematic structure of ultra-cold atoms in a harmonic trap. The red (dark) particles correspond to spin-up fermions and cyan (light) particles corresponds to spin-down fermions.

5.1 Ultra-Cold Fermions in a Harmonic Trap

The first experiments on ultra-cold fermionic atoms in optical lattices focused on non-interacting and attractively interacting particles [79–86]. Only recently, also repulsively interacting fermions in a harmonic trap were studied [89, 90]. Both experiments were performed for two hyperfine states of ^{40}K ($|F, m_F\rangle = |9/2, -9/2\rangle \equiv |\downarrow\rangle$ and $|9/2, -7/2\rangle \equiv |\uparrow\rangle$). To detect the Mott insulator phase the compressibility and double occupancy were studied. These quantities are zero in the Mott insulator phase and finite in a metallic phase. The experiments show that for strongly interacting fermions, in spatial regions with total particle density close to one, a Mott insulator is realized in the system. Depending on the parameters, the Mott insulator can be realized either in the center of the trap, or forms a ring enclosing a metallic and band insulator region. In Ref. [90] it is shown that there is good agreement between the experimental findings and theoretical results obtained by DMFT + TFA calculations.

As it is well known, the repulsive Hubbard model favors antiferromagnetic order at low temperatures. The same type of the behavior is expected in a harmonic trap. At low enough temperatures antiferromagnetic order can be observed by Fourier-sampling of time-of-flight images via Raman pulses, by measuring spin correlation functions via local probes, probing noise correlations, polarization spectroscopy and Bragg scattering.

A repulsively interacting fermionic mixture displays antiferromagnetic spin order at particle density 1. In a harmonic potential, this is only realized in a small region, where the local chemical potential is close to $U_f/2$. Hartree-Fock static mean-field theory predicts that antiferromagnetism, with staggered magnetization on a finite length scale, coexists with paramagnetic states in various spatial patterns, e.g. antiferromagnetism in the center of the trap or antiferromagnetism in a ring surrounded by a particle- or a hole-doped atomic liquid [194]. On the other hand, both commensurate and incommensurate spin-density-waves have been predicted for the hole-doped Hubbard model [195–197]. However, the existence and properties of any ordered state on a finite length scale are strongly sensitive to quantum and thermal fluctuations. Therefore a theoretical description that captures effects of strong correlations and spatial inhomogeneity in a unified framework is needed. We apply the *Real-Space Dynamical Mean-Field Theory* (R-DMFT) [163, 168], which is a comprehensive, thermodynamically consistent and conserving mean-field theory for correlated lattice fermions in the presence of an external inhomogeneous potential. In particular, we apply this method to spin-$\frac{1}{2}$ fermions in a two-dimensional square lattice with harmonic confinement. Here we would like to mention that in our calculations the

5.1 Ultra-Cold Fermions in a Harmonic Trap

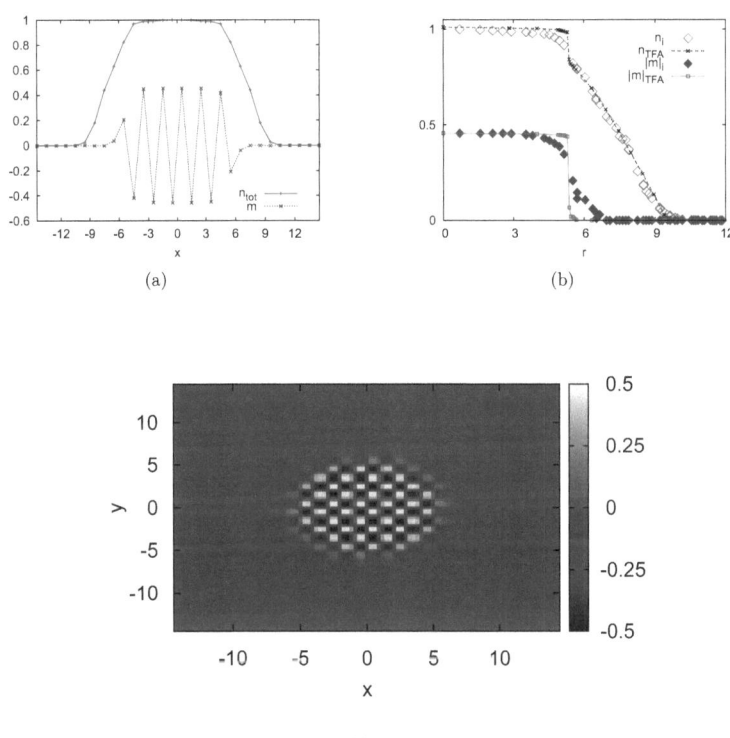

Figure 5.2: Local density and magnetization for $U_f = 10t_f$, $\bar{\mu}_f = 5t_f$ and $V_0^f = 0.1t_f$ on a square (30×30) lattice. In sub-figure (a) we plot the local magnetization and local density along the $y = 1/2$ line (cut through center of the system). In the sub-figure (b) we plot the same values as the function of a radius from the trap center, determined within the exact R-DMFT and within the Thomas-Fermi approximation (TFA) to the R-DMFT. In the sub-figure (c) we show a color-coded plot of the magnetization.

elements of the Green's function that are off-diagonal in spin-space are assumed to be zero. Therefore, we cannot account for canted antiferromagnetism.

Repulsively interacting fermions in an optical lattice almost perfectly implement the Hubbard Hamiltonian

$$\mathcal{H} = -t_f \sum_{\langle ij \rangle, \sigma} \hat{c}^\dagger_{i\sigma} \hat{c}_{j\sigma} + U_f \sum_i \hat{n}_{i\uparrow} \hat{n}_{i\downarrow} + \sum_{i\sigma} (V_i^f - \mu_\sigma) \hat{n}_{i\sigma}, \qquad (5.1)$$

where $\hat{n}_{i\sigma} = \hat{c}^\dagger_{i\sigma} \hat{c}_{i\sigma}$, and $\hat{c}_{i\sigma}$ ($\hat{c}^\dagger_{i\sigma}$) are fermionic annihilation (creation) operators for an atom with spin σ at site i, t_f is the hopping amplitude between nearest neighbor sites $\langle ij \rangle$, $U_f > 0$ is the on-site interaction, μ_σ is the (spin-dependent) chemical potential and $V_i^f = V_0^f r_i^2$ is the harmonic confinement potential. Moreover we define $\bar{\mu}_f \equiv \frac{1}{2}(\mu_\uparrow + \mu_\downarrow)$ and $\Delta\mu \equiv \frac{1}{2}(\mu_\uparrow - \mu_\downarrow)$. The parameters of this model are tunable in experiments by changing the lattice amplitude and via Feshbach resonances. In the following, $t_f = 1$ sets the energy unit and we take the lattice constant to be $a = 1$.

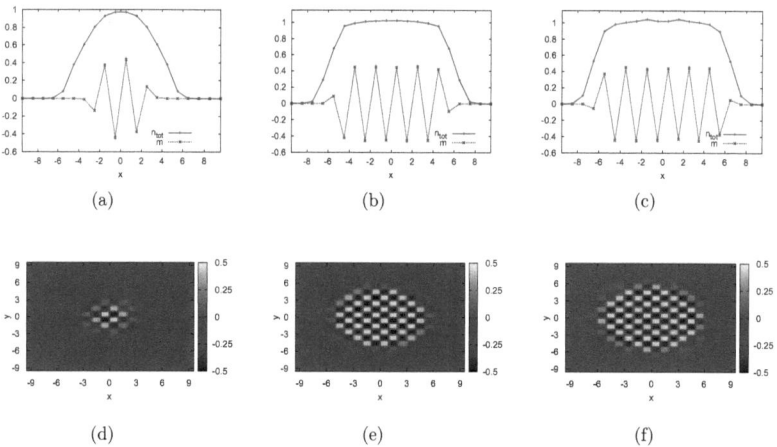

Figure 5.3: The local density and magnetization along the $y = 1/2$ line for $U_f = 10 t_f$ and $V_0^f = 0.2 t_f$ on a square (20×20) lattice. In the sub-figures (a) $\bar{\mu}_f = 3 t_f$, (b) $\bar{\mu}_f = 7 t_f$ and (c) $\bar{\mu}_f = 8 t_f$. In the sub-figures (d), (e), (f) we show a color-coded plot of the magnetization for the same parameter as in (a),(b) and (c) correspondingly.

In the context of cold atoms, a two dimensional system can be realized by applying

5.1 Ultra-Cold Fermions in a Harmonic Trap

a highly anisotropic optical lattice, which divides the system into two-dimensional slices. Although not exact, the R-DMFT is expected to be a good approximation for the two-dimensional situation at zero temperature, since the derivation of the DMFT equations is controlled by the small parameter $1/z = 1/4$ on the square lattice.

We show that for spin-$\frac{1}{2}$ lattice fermions with local repulsive interaction, antiferromagnetic order exists at zero temperature when the harmonic potential is present. We find that antiferromagnetic order is stable in spatial regions with total particle density close to one, but persists also in parts of the system where the local density significantly deviates from half filling. We also show that for strong repulsion phase separation occurs in imbalanced mixtures, when the difference in the particle number of the spin components is large. For weaker repulsion a strong imbalance destroys antiferromagnetic order, but does not lead to phase separation.

5.1.1 Balanced Mixture

In this subsection we consider the case of an equal mixture of spin-up and down atoms: $N_\uparrow = N_\downarrow$. First we consider the case when the fermions are half-filled at the center of trap, i.e. $\bar{\mu}_f = U_f/2$. In this case our calculations show that *Antiferromagnetic* (AF) order appears in the center of the system (Fig. 5.2).

To investigate the effect of the number of fermions on the behavior of the system we change the chemical potential. We obtain that in a wide range of chemical potentials antiferromagnetic order is stable in the center of the trap (Fig. 5.3), but for high chemical potentials, when the local density of the fermions in the center of the system becomes much higher than half-filling, antiferromagnetic order forms in a ring enclosing a paramagnetic region (Fig. 5.4).

So we observe that antiferromagnetic order is stable in the presence of the inhomogeneous harmonic potential. These results are particularly important for ongoing attempts to realize antiferromagnetic states in optical lattices. Namely, we predict that the observation of antiferromagnetic order does not critically depend on the number of atoms in the system. For sufficiently strong repulsion between the particles, the necessary condition to find antiferromagnetic order is to prepare the system such that the local filling factor approximates or exceeds one in at least part of the system. We find no evidence for phase separation or a paramagnetic insulating boundary layer for the $N_\uparrow = N_\downarrow$ case.

The antiferromagnetic ground state of homogeneous fermions described by the Hubbard Hamiltonian (5.1) without a trap is stable when the density of particles varies from $n \approx 0.8$

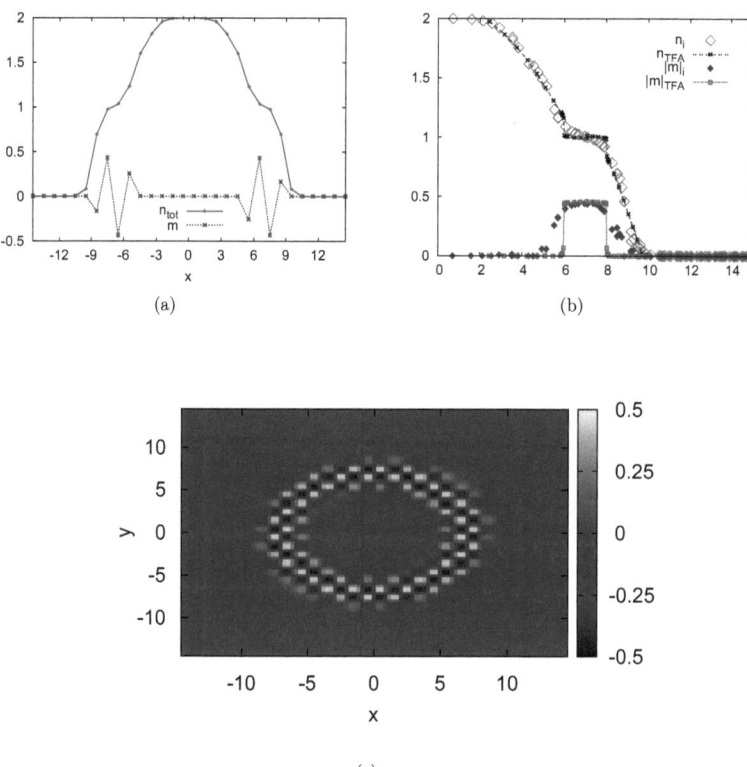

Figure 5.4: Local density and magnetization for $U_f = 10t_f$, $\bar{\mu}_f = 15t_f$ and $V_0^f = 0.2t_f$ on a square (30×30) lattice. In sub-figure (a) we plot the local magnetization and local density along the $y = 1/2$ line. In the sub-figure (b) we plot the same values as a function of the radius from the trap center, determined within the exact R-DMFT and within the Thomas-Fermi approximation (TFA) to the R-DMFT. And in the sub-figure (c) we show color-coded plot for the magnetization.

5.1 Ultra-Cold Fermions in a Harmonic Trap

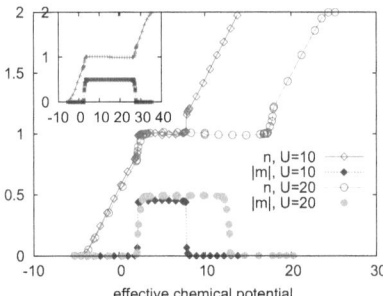

Figure 5.5: Total density and staggered magnetization as a function of the effective chemical potential obtained within the Thomas-Fermi approximation to the R-DMFT for two dimensional square (main panel) and three dimensional cubic lattices (inset). Main panel: $U_f = 10 t_f$ (diamonds) and $U_f = 20 t_f$ (circles); inset: $U_f = 30 t_f$.

to 1.2, depending on the interaction value U_f [198]. On the contrary, in the presence of the external harmonic potential, antiferromagnetic order appears for much lower or higher local total densities. Indeed, in Figs. 5.2(b) and 5.4(b) we present examples of the local density n_i and the local magnetization $m_i = \frac{1}{2}(n_{i\uparrow} - n_{i\downarrow})$ as a function of distance from the center which prove that antiferromagnetic order extends from the center of the trap and disappears only when $n_i \approx 0.5$ in Fig. 5.2(b). Similarly, Fig. 5.4(b) shows that antiferromagnetic order is stable on a ring when the local density extends between $0.5 \lesssim n_i \lesssim 1.5$.

We also determine the local density and the local magnetization within the Thomas Fermi approximation (TFA) to the R-DMFT, where the external potential is only included by a spatially varying chemical potential [199]. The agreement between the full R-DMFT and the TFA results is very good in regions well within or outside the antiferromagnetic domain (see Figs. 5.2(b) and 5.4(b)). Encouraged by this, Fig. 5.5 shows additional TFA+R-DMFT profiles that can be used to compare R-DMFT with experiments for realistic systems in two and three dimensions. However, the staggered magnetization decays abruptly within the TFA as compared to the full R-DMFT solution, i.e. the Thomas-Fermi approximation to the R-DMFT essentially reproduces results from the standard homogeneous DMFT, cf. Fig. 5.5. The wider stability regime of antiferromagnetic order found within the full R-DMFT is caused by a proximity effect; antiferromagnetic order is induced into parts of the systems where the local densities are too low to stabilize antiferromagnetism in the homogeneous case. On the other hand, the proximity of the paramagnetic state

78   5. Ultra-Cold Atoms in a Harmonic Trap

reduces the staggered magnetization when the local density is close to one.

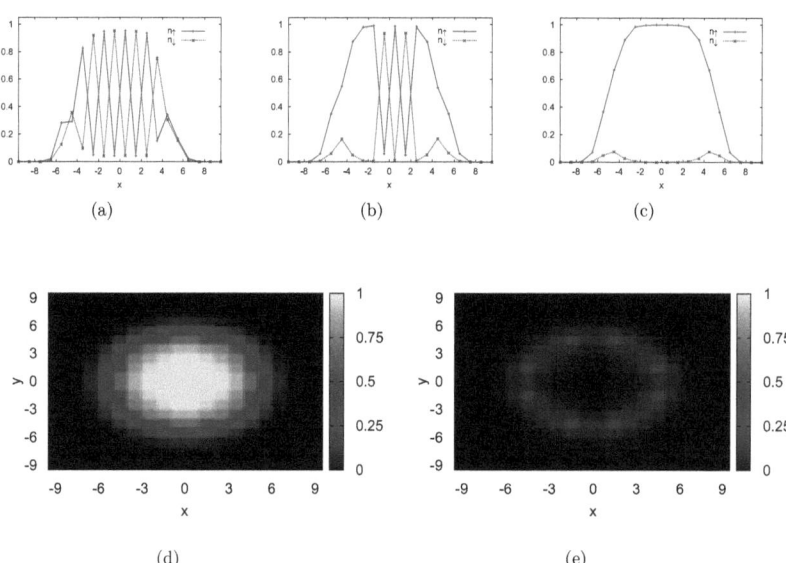

Figure 5.6: Spin resolved particle densities for an imbalanced mixture obtained within the R-DMFT for $U_f = 10t_f$. Panels (a)-(c) show component densities along the $y = \frac{1}{2}$ line for gradually increasing imbalanced $\Delta\mu = 0.3t_f$ (a), $0.75t_f$ (b), t_f (c). The two lower panels show the space resolved up- (d) and down- (e) density for $\Delta\mu = t_f$. The lattice size is 20×20 and other parameters are: $V_0^f = 0.2t_f$, $\bar{\mu} = 5t_f$.

5.1.2 Imbalanced Mixture

We now proceed by investigating the imbalanced case, i.e. $N_\uparrow \neq N_\downarrow$. Imbalance between the two spin-components is induced by a nonzero chemical potential difference $\Delta\mu = \mu_\uparrow - \mu_\downarrow$, which corresponds to a magnetic field. In the experimental situation, the density imbalance can be highly tuned and is stable due to the suppression of spin-flip scattering processes in cold-atomic gases. Representative results are presented in Fig. 5.6, where we plot the up- and down-component of the density along a cut through the system. Upon increasing the imbalance parameter $\Delta\mu$, we find suppression of antiferromagnetic order and emergence of

5.1 Ultra-Cold Fermions in a Harmonic Trap

phase separation between the minority and majority species. The phase separation region starts to develop far away from the center of the trap at small $\Delta\mu$ and gradually spreads toward the center. We thus find that the border of the antiferromagnetic domain is most sensitive to phase separation. This is indeed reasonable: the energy cost to polarize the antiferromagnetic state is the energy difference between the antiferromagnetic state and the ferromagnetic state. This is of the order t_f^2/U_f, which is small for the large interaction U_f considered here. Antiferromagnetic order is thereby more unstable for larger distances to the trap-center, because of the vicinity to the paramagnetic region. The energy cost to polarize the paramagnetic regime is higher, because in this case kinetic energy has to be paid, whereas in the antiferromagnetic domain the kinetic energy is already quenched because the particles are almost localized. Due to the proximity effect we find that the paramagnetic regime close to the insulating domain also gets phase-separated, which leads to a ring-like structure of the minority species.

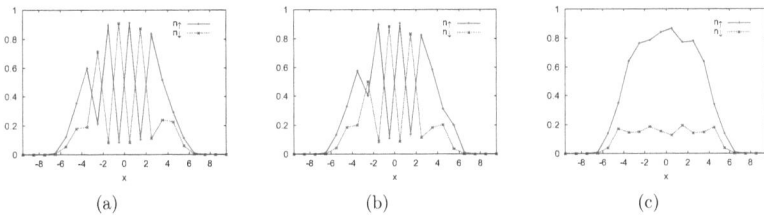

Figure 5.7: Spin resolved particle densities for an imbalanced mixture obtained within the R-DMFT for $U_f = 7.5 t_f$. Panels (a)-(c) show component densities along the $y - \frac{1}{2}$ line for gradually increasing imbalanced $\Delta\mu = 0.4 t_f$ (a), $0.6 t_f$ (b), $0.8 t_f$ (c). The lattice size is 20×20 and other parameters are: $V_0^f = 0.2 t_f$, $\bar{\mu} = 3.75 t_f$.

At strong interaction, $U_f = 10 t_f$, in the case shown in Fig. 5.6, atoms with different spins ultimately tend to occupy different spatial regions to avoid the mutual interaction and the minority species is completely expelled from the trap center. At weaker interaction, $U_f = 7.5_f$, however, we found that the imbalanced system still contains interpenetrating atoms with different spins and phase separation does not occur. This is shown in Fig. 5.7, where for $\Delta\mu = 0.8 t_f$ antiferromagnetic order has completely disappeared, but the two spin components are still interpenetrating. The small oscillations in the component densities can be understood as Friedel oscillations due to the small size of the system. We note that in the case of imbalanced spin-mixtures the agreement between the TFA and the exact solution to R-DMFT is far less good than in the balanced case presented above.

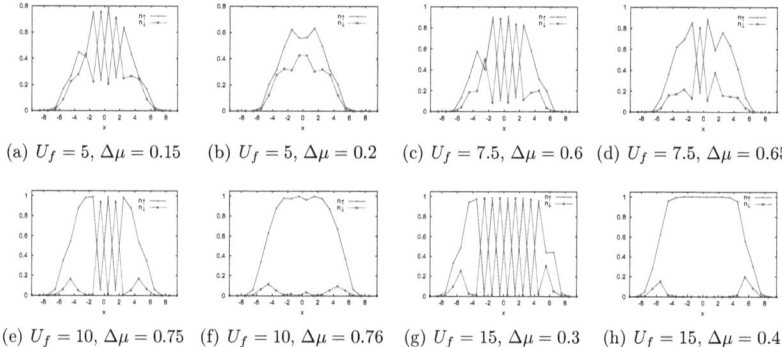

(a) $U_f = 5$, $\Delta\mu = 0.15$ (b) $U_f = 5$, $\Delta\mu = 0.2$ (c) $U_f = 7.5$, $\Delta\mu = 0.6$ (d) $U_f = 7.5$, $\Delta\mu = 0.65$

(e) $U_f = 10$, $\Delta\mu = 0.75$ (f) $U_f = 10$, $\Delta\mu = 0.76$ (g) $U_f = 15$, $\Delta\mu = 0.3$ (h) $U_f = 15$, $\Delta\mu = 0.4$

Figure 5.8: Spin resolved particle densities for an imbalanced mixture obtained within the R-DMFT for different values of U_f and $\Delta\mu$. $\bar{\mu}_f = U_f/2$ and $V_0^f = 0.2 t_f$.

To understand better the effect of the imbalance on antiferromagnetic order we also investigate other values of interaction strength. Our results show that for $U_f < U_{\mathrm{max}}$ the critical imbalance $\Delta\mu_c$, which breaks antiferromagnetic order, is increasing with increasing of U_f, while for $U_f > U_{\mathrm{max}}$ the critical imbalance $\Delta\mu_c$ is decreasing with increasing of U_f (see Fig. 5.8). The reason of a such behavior is the following: for weak interaction U_f antiferromagnetic order is of *Slater type* (spin waves) and the gap is linear in U_f, while in the strong interaction U_f antiferromagnetic order is of *Heisenberg type* (localized moments) and the gap is order of t_f^2/U_f. So, with increasing interaction U_f the size of the gap is increasing, reaches its maximum and than decreasing and approaches to zero. This fully explains the behavior of the critical imbalance as a function of U_f.

5.2 Mixtures of Spinless Fermions and Bosons in a Harmonic Trap

In chapter 4, we considered a mixture of the spinless fermions and bosons in a homogeneous optical lattice. We showed that when the fermions are half-filled, depending on the model parameters one can obtain the supersolid or alternative Mott insulator (AMI) phase. In this section we will consider the effect of the trap on the Bose-Fermi mixture.

A Bose-Fermi mixture in an optical lattice is well described by the Bose-Fermi Hubbard model:

$$\mathcal{H} = -t_f \sum_{\langle ij \rangle} \hat{c}_i^\dagger \hat{c}_j - t_b \sum_{\langle ij \rangle} \hat{b}_i^\dagger \hat{b}_j + U_b \sum_i \hat{n}_i^b(\hat{n}_i^b - 1) + U_{fb} \sum_i \hat{n}_i^b \hat{n}_i^f + \sum_{i\sigma}(V_i^f - \mu_f)\hat{n}_i^f + \sum_{i\sigma}(V_i^b - \mu_b)\hat{n}_i^b, \quad (5.2)$$

where $\hat{c}_i^{(\dagger)}$ and $\hat{b}_i^{(\dagger)}$ are annihilation (creation) operators for fermions and bosons at site i. $\hat{n}_i^f = \hat{c}_i^\dagger \hat{c}_i$ and $\hat{n}_i^b = \hat{b}_i^\dagger \hat{b}_i$ are the number operators for fermions and bosons at site i respectively. $t_{f(b)}$ is the fermionic (bosonic) hopping amplitude between nearest neighbor sites $\langle ij \rangle$. U_b and U_{fb} are the Bose-Bose and Bose-Fermi interactions, respectively. $\mu_{f(b)}$ is the fermionic (bosonic) chemical potential and $V_i^{f(b)} = V_0^{f(b)} r_i^2$ is the harmonic confinement potential for fermions (bosons).

First we consider a mixture of spinless fermions and hard-core bosons in the harmonic trap. We choose the chemical potentials in such a way that in the center of the trap both fermions and bosons are half-filled. We take hopping amplitudes and harmonic confinement potential for fermions and bosons equal to each other. Our calculations show that a ring-like structure is realized in the system. In the center we obtain an AMI phase, surrounded by the supersolid and superfluid phases (see Fig. 5.9). The supersolid phase obtained during this calculations is not a "true" supersolid and it arises due to boundary effects. Without the AMI phase in the center of the trap we can not obtain a "true" supersolid phase with CDW amplitude less than 1/2. This behavior can be explained by the fact that the gap in the supersolid phase, as we discussed in chapter 4, is rather small, and is easily affected by the harmonic confinement, while the gap in the AMI phase is larger and more stable.

As we discussed in section 4.1.2 the gap in the supersolid phase is increasing with increasing filling of the bosons. Therefor, to obtain a supersolid phase we chose a bosonic chemical potential, such that filling in the center of the trap is 3/2. Our results are summarized in Fig. 5.10. We again obtain a ring-like structure: in the center we obtain a

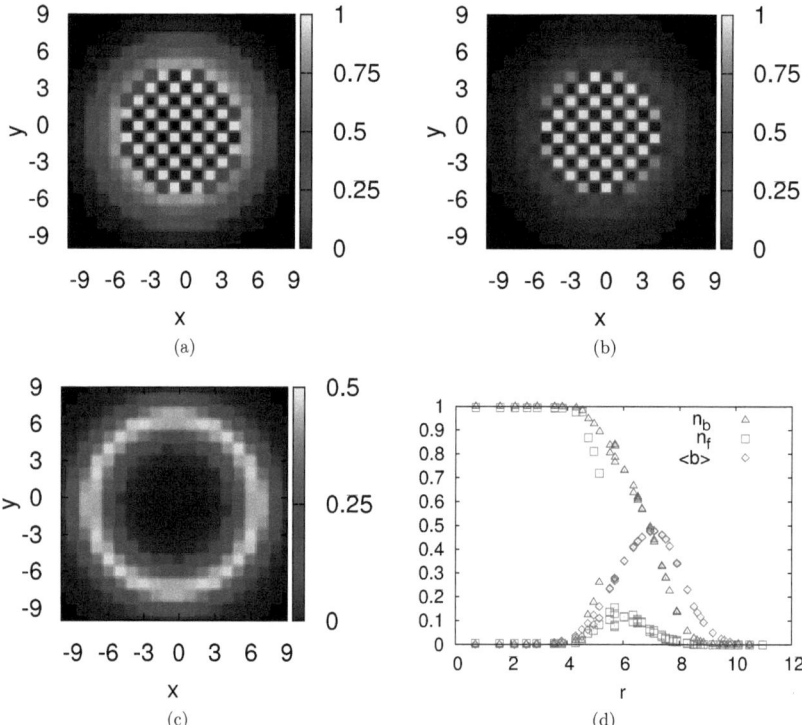

Figure 5.9: Local density for hard-core bosons and fermions, as well as the superfluid order parameter for $t_b = t_f$, $V_f = V_b = 0.2t_f$, $\mu_f = 10t_f$, $\mu_b = 10t_b$ and $U_{fb} = 20t_f$ on a square (20×20) lattice. In sub-figure (a), (b) and (c) we represent a color-coded plots for local densities for bosons and fermions, and superfluid order parameter, respectively. In the sub-figure (d) we plot the same values as the function of the radius from the trap center.

5.2 Mixtures of Spinless Fermions and Bosons in a Harmonic Trap   83

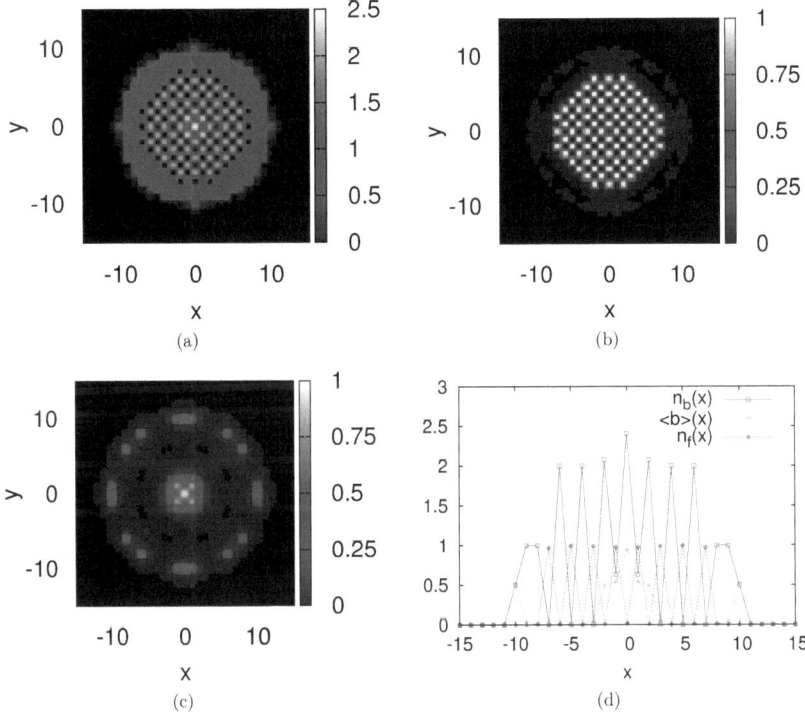

Figure 5.10: Local density for bosons and fermions, as well as the superfluid parameter for $t_b = 0.05t_f$, $V_f = 0.16t_f$, $V_b = 0.08t_f$, $\mu_f = 12t_f$, $\mu_b = 160t_b$, $U_{fb} = 8t_f$, and $U_b = 80t_b$ on a square (31×31) lattice. In sub-figure (a), (b) and (c) we represent color-coded plots for local densities for bosons and fermions, and superfluid order parameter, respectively. In the sub-figure (d) we plot the same quantities along the $y = 0$ line.

supersolid surrounded by AMI and superfluid phases.

In conclusion, we have shown that the supersolid and AMI phases found in Chapter 3 can be stabilized in an harmonic trap. However, the supersolid is very sensitive and it's existence is easily destroyed by the trap, because of the small gap in the spectral function. The AMI phase, on the other hand, can be stabilized in the harmonic potential for a wide range of couplings.

Chapter 6

Resonance Superfluidity in an Optical Lattice

In this chapter we study an ultracold atomic gas of fermionic atoms in a three-dimensional optical lattice close to a Feshbach resonance. For magnetic field values below the resonance, fermions with different spin will form bosonic molecules (See Fig. 6.1). Varying the magnetic field one can detune the bosonic level compared to the fermionic one. Doing this one can vary the ratio of the filling of fermions and molecular bosons and the effective interaction between the fermions. In such a system interesting physics can be observed. In particular, the Feshbach resonance induces a BEC-BCS crossover close to resonance position. Far away from the resonance the behavior of the system is dominated by the background scattering length.

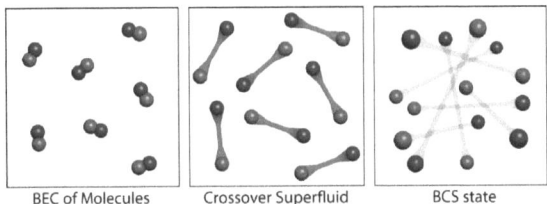

Figure 6.1: The BEC-BCS crossover. By tuning the interaction strength between the two fermionic spin states, one can smoothly cross over from a regime of tightly bound molecules to a regime of long-range Cooper pairs, whose characteristic size is much larger than the interparticle spacing. In between these two extremes, one encounters an intermediate regime where the pair size is comparable to the interparticle spacing (From Ref. [95]).

T. Esslinger and colleagues at ETH Zürich reported the production of ^{40}K molecules in 3D cubic optical lattices using s-wave Feshbach resonances in early 2006 [84], but no evidence of a superfluid state was found until later that year, when Ketterle and coworkers loaded ^{6}Li atoms in optical lattices and their pairs formed a condensate [85]. Those two experiments opened the door to studies of superfluid-to-insulator transitions in optical lattices.

6.1 Microscopic Model

Studying ultracold fermions close to a Feshbach resonance is a challenging problem. Due to the fact that exactly on resonance the scattering length passes through $\pm\infty$ the Hamiltonian in Eq. (2.42) cannot be defined. To deal with this problem, it is necessary to formulate a Hamiltonian by separating out the resonance state and treating it explicitly [200]. The nonresonant contributions give rise to a background scattering length and characterize interactions between fermions. As the Feshbach resonance occurs due to a coupling with the bosonic molecular state, one has to explicitly introduce bosonic degrees of freedom in the theory to describe resonant processes [200].

An ultracold atomic gas of fermionic atoms and bosonic molecules close to a Feshbach resonance in the presence of an optical lattice is well described by a Bose-Fermi Hubbard model [201, 202]. In our calculation we assume the bosons to be in the lowest band. For the fermions, on the other hand, we have to take into account also the higher bands, since the Feshbach problem in an optical lattice is inherently a multi-band problem [202, 203]. Since the bandwidth is much smaller than the band gap, we approximate the higher bands to be flat and only take into account the full band-structure for the lowest band. Moreover, we neglect the interaction between fermions in higher bands with each other and with the bosons. This is justified because the filling in the higher bands is very small, such that

6.1 Microscopic Model

interaction effects are also small. The Hamiltonian thus has the following form:

$$\hat{\mathcal{H}} = \hat{\mathcal{H}}_f^0 + \hat{\mathcal{H}}_b + \hat{\mathcal{H}}_{fb}^0 + \sum_{l=1}^{\infty}(\hat{\mathcal{H}}_f^l + \hat{\mathcal{H}}_{fb}^l)\,, \qquad (6.1)$$

$$\hat{\mathcal{H}}_f^0 = -t_f \sum_{\langle i \rangle} \hat{c}_{i\sigma,0}^\dagger \hat{c}_{j\sigma,0} + U_f \sum_i \hat{n}_{i,\uparrow,0}^f \hat{n}_{i,\downarrow,0}^f - (\mu - \frac{3\hbar\omega}{2}) \sum_i \hat{n}_{i,0}^f\,, \qquad (6.2)$$

$$\hat{\mathcal{H}}_b = -t_b \sum_{\langle i \rangle} \hat{b}_i^\dagger \hat{b}_j + \frac{U_b}{2} \sum_i \hat{n}_i^b(\hat{n}_i^b - 1) - (2\mu - \delta - \frac{3\hbar\omega}{2}) \hat{n}_i^b\,, \qquad (6.3)$$

$$\hat{\mathcal{H}}_{fb}^0 = U_{fb} \sum_i \hat{n}_i^b \hat{n}_{i,0}^f + g_0 \sum_i \left(\hat{b}_i^\dagger \hat{c}_{i\uparrow,0} \hat{c}_{i\downarrow,0} + h.c\right)\,, \qquad (6.4)$$

$$\hat{\mathcal{H}}_f^l = \sum_i \left(\left(2l + \frac{3}{2}\right)\hbar\omega - \mu\right) \hat{n}_{i,l}^f\,, \qquad (6.5)$$

$$\hat{\mathcal{H}}_{fb}^l = g_l \sum_i \left(\hat{b}_i^\dagger \hat{c}_{i\uparrow,l} \hat{c}_{i\downarrow,l} + h.c\right)\,, \qquad (6.6)$$

where $\hat{c}_{i\sigma,l}^\dagger$ is the creation operator of a fermion with spin σ for the l-th band on lattice site i. $\hat{b}_i^\dagger$ is the creation operator of a boson at site i. $\hat{n}_{i\sigma,l}^f = \hat{c}_{i\sigma,l}^\dagger \hat{c}_{i\sigma,l}$, $\hat{n}_{i,l}^f = \hat{n}_{i\uparrow,l}^f + \hat{n}_{i\downarrow,l}^f$ are the fermionic number operators, and $\hat{n}_i^b = \hat{b}_i^\dagger \hat{b}_i$ is the bosonic number operator. U_f, U_b and U_{fb}, are the on-site Hubbard interactions between the fermions, bosons and fermions and bosons in the lowest band, respectively. μ is the chemical potential. δ is the detuning of the bosonic level, ω is the frequency of the harmonic oscillator associated with an optical lattice minimum. $g_l = g_0\sqrt{L_l^{(1/2)}(0)}$ is the Feshbach coupling, where g_0 is the Feshbach coupling for the lowest Hubbard band and $L_l^{(1/2)}(0)$ is the generalized Laguerre polynomial [203].

$$t_{b(f)} \simeq \frac{4}{\sqrt{\pi}} E_r^{b(f)} \left(\frac{V_0}{E_r^{b(f)}}\right)^{3/4} \exp\left[-2\sqrt{\frac{V_0}{E_r^{b(f)}}}\right]\,, \qquad (6.7)$$

$$U_{b(f)} \simeq \sqrt{\frac{8}{\pi}} k a_{b(f)} E_r^{b(f)} \left(\frac{V_0}{E_r^{b(f)}}\right)^{3/4}\,, \qquad (6.8)$$

$$U_{fb} \simeq \frac{4}{\sqrt{\pi}} k a_{fb} E_r^b \frac{1 + m_b/m_f}{(1 + \sqrt{m_b/m_f})^{3/2}} \left(\frac{V_0}{E_r^b}\right)^{3/4}\,, \qquad (6.9)$$

$$g_0 = \hbar\sqrt{\frac{4\pi a_f \Delta B \Delta \mu_{\text{mag}}}{m_f}} \left(\frac{m_f \omega}{2\pi\hbar}\right)^{3/4}\,, \qquad (6.10)$$

$$\delta = \Delta\mu_{\text{mag}}(B - B_0)\,. \qquad (6.11)$$

Here a_f, a_b, and a_{fb} are fermion-fermion, boson-boson, and fermion-boson background scattering lengths. Here we approximate the background boson-boson scattering length and Bose-Fermi scattering by $a_b = 0.6a_f$ [109] and $a_{bf} = 1.2a_f$ [204]. Furthermore, B is the magnetic field, and B_0 and ΔB are the position of the Feshbach resonance and its width, respectively. $\Delta \mu_{\text{mag}}$ is the difference in the magnetic moment between the closed and open channel of the Feshbach resonance. Finally, m_f and m_b are the respective masses of the fermions and bosons.

One can simplify the Hamiltonian (6.1) by the following rescaling:

$$\bar{\mu} = \mu - \frac{3\hbar\omega}{2}, \qquad (6.12)$$

$$\bar{\delta} = \delta - \frac{3\hbar\omega}{2}. \qquad (6.13)$$

After this rescaling the factor $\frac{3\hbar\omega}{2}$ disappears.

6.2 Method

The multi-band Hamiltonian derived so far is very complicated, since it both involves strong correlations and many bands. Simply neglecting the higher bands, leads to a wrong description close to the Feshbach resonance, since the Feshbach parameter g is very large there, even exceeding the band gap [202]. However, the filling of fermions in the higher bands is strongly suppressed by the band gap. This allows us to make a mean-field decoupling in the higher bands [202]. The lowest band is left untouched in this procedure, since the fermionic filling can be very large there.

We thus make the following decoupling for $l > 0$:

$$\hat{\mathcal{H}}_{fb}^{li} = g_l \left(\langle \hat{b}_i^\dagger \rangle \hat{c}_{i\uparrow,l} \hat{c}_{i\downarrow,l} + \hat{b}_i^\dagger \langle \hat{c}_{i\uparrow,l} \hat{c}_{i\downarrow,l} \rangle + h.c \right). \qquad (6.14)$$

This step implies that the lowest band and the higher bands are only coupled in a mean-field way. They can thus be diagonalized separately, but the coupling arises because of the mean-field self-consistency relations. The full Hamiltonian is now given by

$$\hat{\mathcal{H}} = \hat{\mathcal{H}}_f^0 + \hat{\mathcal{H}}_b + \hat{\mathcal{H}}_{fb}^0 + \sum_i \hat{\mathcal{H}}_b'(i), \qquad (6.15)$$

6.2 Method

where the following terms are added to the bosonic part of the lowest band Hamiltonian:

$$\hat{\mathcal{H}}'_b(i) = \sum_{l=1} g_l \left(\hat{b}^\dagger_i \langle \hat{c}_{i\uparrow,l} \hat{c}_{i\downarrow,l} \rangle + h.c \right) = -\left(\Delta \hat{b}^\dagger_i + h.c. \right). \qquad (6.16)$$

For each of the higher bands $l > 0$ we obtain the Hamiltonian (here we suppress the site index i):

$$\hat{\mathcal{H}}^l_f = \begin{pmatrix} \hat{c}^\dagger_{l\uparrow} \\ \hat{c}_{l\downarrow} \end{pmatrix} \begin{pmatrix} 2l\hbar\omega - \bar{\mu} & g_l \langle \hat{b} \rangle \\ g_l \langle \hat{b}^\dagger \rangle & -(2l\hbar\omega - \bar{\mu}) \end{pmatrix} \begin{pmatrix} \hat{c}_{l\uparrow} \\ \hat{c}^\dagger_{l\downarrow} \end{pmatrix}. \qquad (6.17)$$

One has to solve this problem self-consistently.

The solution of Eq. (6.17) is well known and has the following form:

$$\omega_l = \sqrt{(2l\hbar\omega - \bar{\mu})^2 + g_l^2 |\langle \hat{b} \rangle|^2}, \qquad (6.18)$$

$$u_l^2 = \frac{1}{2} + \frac{2l\hbar\omega - \bar{\mu}}{2\omega_l}, \qquad (6.19)$$

$$v_l^2 = \frac{1}{2} - \frac{2l\hbar\omega - \bar{\mu}}{2\omega_l}, \qquad (6.20)$$

$$u_l v_l = \frac{g_l \langle \hat{b} \rangle}{2\omega_l}. \qquad (6.21)$$

So we obtain that:

$$n_{fl} = 2v_l^2 + 2(u^2 - v^2)f(\omega_l) = 1 - \frac{2l\hbar\omega - \bar{\mu}}{\omega_l} \tanh\left(\frac{\omega_l}{2kT}\right), \qquad (6.22)$$

$$|\langle \hat{c}_{l\uparrow} \hat{c}_{l\downarrow} \rangle| = |u_l v_l| \tanh\left(\frac{\omega_l}{2kT}\right) = \left| \frac{g_l \langle \hat{b} \rangle}{2\omega_l} \right| \tanh\left(\frac{\omega_l}{2kT}\right), \qquad (6.23)$$

where $f(\omega_l)$ is the Fermi function and T is the temperature. We used absolute values in the equation for $\langle \hat{c}_{l\uparrow} \hat{c}_{l\downarrow} \rangle$, because of the ambiguity of the sign, coming from the fact that still a divergence has to be subtracted (see below).

The total number of fermions is now equal to:

$$n_{ftot} = n_{f0} + \sum_{l=1}^{\infty} \left(1 - \frac{2l\hbar\omega - \bar{\mu}}{\omega_l} \tanh\left(\frac{\omega_l}{2kT}\right) \right). \qquad (6.24)$$

This is a converging sum, which can be evaluated numerically.

From Eq. (6.16) follows that we have to evaluate the sum:

$$\sum_{l=1} g_l \langle \hat{c}_{l\uparrow} \hat{c}_{l\downarrow} \rangle = \pm \langle \hat{b} \rangle \sum_{l=1} \frac{g_l^2}{2\omega_l} \tanh\left(\frac{\omega_l}{2kT}\right). \quad (6.25)$$

This sum is diverging. This is the physical divergence that always arises when approximating the T-matrix by a delta-potential [202, 203]. This problem can be solved by using a pseudo-potential [203]. Here we follow the way of Ref. [202] and isolate the diverging contribution from the problem.

First, we notice that for large l, ω_l can be approximated by $\omega_l = 2l\hbar\omega - \bar{\mu}$ and one can also approximate $\tanh\left(\frac{\omega_l}{2kT}\right)$ by one. So

$$\sum_{l=1} \frac{g_l^2}{2\omega_l} \tanh\left(\frac{\omega_l}{2kT}\right) \simeq \left(\sum_{l=1}^{N} \frac{g_l^2}{2\omega_l} \tanh\left(\frac{\omega_l}{2kT}\right) + \sum_{l=N+1}^{\infty} \frac{g_l^2}{2(2l\hbar\omega - \bar{\mu})} \right)$$
$$= \left(\sum_{l=1}^{N} \frac{g_l^2}{2\omega_l} \tanh\left(\frac{\omega_l}{2kT}\right) - \sum_{l=0}^{N} \frac{g_l^2}{2(2l\hbar\omega - \bar{\mu})} + \sum_{l=0}^{\infty} \frac{g_l^2}{2(2l\hbar\omega - \bar{\mu})} \right). \quad (6.26)$$

Here N is a large integer number (in our calculation we took $N = 500$).

The first two terms of Eq. (6.26) are finite sums, but the last term is diverges. This sum is known from the literature [203, 205]. To separate the diverging part, we have to take the following limit:

$$\sum_{l=0}^{\infty} \frac{g_l^2}{2(2l\hbar\omega - \bar{\mu})} = \sum_{l=0}^{\infty} \frac{g_0^2 L_l^{(1/2)}(0)}{2(2l\hbar\omega - \bar{\mu})} = \lim_{r\to 0} \sum_{l=0}^{\infty} \frac{g_0^2 L_l^{(1/2)}(r)}{2(2l\hbar\omega - \bar{\mu})} \quad (6.27)$$
$$= -\lim_{r\to 0} \left(\frac{g_0^2 \sqrt{\pi} \Gamma(-\bar{\mu}/2\hbar\omega)/\Gamma(-\bar{\mu}/2\hbar\omega - 1/2)}{2\hbar\omega} - \frac{\sqrt{\pi}}{r} + \mathcal{O}(r) \right).$$

Since the diverging part does not depend on the model parameters, we can cure the divergence by neglecting this term [202, 203]. Doing so, we obtain

$$\Delta = -\sum_{l=1}^{\infty} g_l \langle \hat{c}_{l\uparrow} \hat{c}_{l\downarrow} \rangle = \pm \langle \hat{b} \rangle \left(\frac{g_0^2 \sqrt{\pi} \Gamma(-\bar{\mu}/\hbar\omega)/\Gamma(-\bar{\mu}/\hbar\omega - 1/2)}{\hbar\omega} \right.$$
$$\left. + \sum_{l=0}^{N} \frac{g_l^2}{2(2l\hbar\omega - \bar{\mu})} - \sum_{l=1}^{N} \frac{g_l^2}{2\omega_l} \tanh\left(\frac{\omega_l}{2kT}\right) \right). \quad (6.28)$$

We now fix the sign by requiring $\Delta > 0$. The reason is that this solution minimizes the (free) energy.

6.2 Method

Summarizing, we have reduced the multi-band problem to a single band problem:

$$\hat{\mathcal{H}} = \hat{\mathcal{H}}_f^0 + \hat{\mathcal{H}}_b + \hat{\mathcal{H}}_b' + \hat{\mathcal{H}}_{fb}^0, \quad (6.29)$$

where $\hat{\mathcal{H}}_f^0$, $\hat{\mathcal{H}}_b$, $\hat{\mathcal{H}}_{fb}^0$ and $\hat{\mathcal{H}}_b'$ are described by the Eqs. (6.2), (6.3), (6.4) and (6.16), respectively.

The chemical potential μ has to be adjusted, such that the total filling is equal to the desired value n_{tot}:

$$n_b^0 + n_f^0 + \sum_{l=1}^{\infty} \left(1 - \frac{2l\hbar\omega - \bar{\mu}}{\omega_l} \tanh\left(\frac{\omega_l}{2kT}\right)\right) = n_{tot}. \quad (6.30)$$

This leads to the following self-consistency loop: we start from an initial guess of the superfluid order parameter $\langle \hat{b} \rangle$ and using Eq. (6.28) we calculate Δ. This means that we know all parameters in the Hamiltonian (6.29), and can find its eigenvalues and eigenvectors, and correspondingly calculate new correlation functions, including the superfluid order parameter $\langle \hat{b} \rangle$. With this step the self-consistency loop is closed.

To deal with the Hamiltonian (6.29) we use generalize dynamical mean field theory (GDMFT) [20, 21]. The GDMFT method is explained in detail in section 3.2. In this case the system is described by the following Hamiltonian:

$$\hat{\mathcal{H}} = \hat{\mathcal{H}}_b + \hat{\mathcal{H}}_{fb} + \hat{\mathcal{H}}_f, \quad (6.31)$$

$$\hat{\mathcal{H}}_b = -\left[(zt_b\varphi + \Delta)\hat{b}^\dagger + h.c.\right] + \frac{U_b}{2}\hat{n}^b(\hat{n}^b - 1) - \mu_b \hat{n}^b,$$

$$\hat{\mathcal{H}}_{fb} = U_{fb}\hat{n}^f \hat{n}^b + g_0\left(\hat{b}_i^\dagger \hat{c}_{i\uparrow,0}\hat{c}_{i\downarrow,0} + h.c\right),$$

$$\hat{\mathcal{H}}_f = -\mu_{\sigma f}\hat{n}^f + U_f \hat{n}_\uparrow^f \hat{n}_\downarrow^f + W_l \left(\hat{a}_{l\uparrow}^\dagger \hat{a}_{l\downarrow}^\dagger + h.c.\right)\}$$

$$+ \sum_{l,\sigma}\left\{\varepsilon_{l\sigma}\hat{a}_{l\sigma}^\dagger \hat{a}_{l\sigma} + V_{l\sigma}\left(\hat{c}_\sigma^\dagger \hat{a}_{l\sigma} + h.c.\right),\right.$$

Here z is the lattice coordination number, $\varphi = \langle \hat{b} \rangle$ is the superfluid order parameter. l labels the noninteracting orbitals of the effective bath, $V_{l\sigma}$ are the corresponding fermionic hybridization matrix elements and W_l describes superconducting propaties of the bath.

The combination of the mean-field approximation in the higher bands and the GDMFT, however, leads to a problem. Both approximations couple to the superfluid order parameter $\langle \hat{b}_i \rangle$. The mean-field approximation for the higher bands, means that the local correlator $\langle \hat{b}_i^\dagger \hat{c}_{i\uparrow,l} \hat{c}_{i\downarrow,l} \rangle$ is approximated by $\langle \hat{b}_i^\dagger \rangle \langle \hat{c}_{i\uparrow,l} \hat{c}_{i\downarrow,l} \rangle$. The GDMFT scheme, on the other hand, involves the approximation to replace the non-local correlator $\langle \hat{b}_i^\dagger \hat{b}_j \rangle$ by $\langle \hat{b}_i^\dagger \rangle \langle \hat{b}_j \rangle$. This means that $\langle \hat{b} \rangle$ both measures the local phase coherence between the bosons and fermions and the non-local bosonic long range order. However, these are two very different quantities and generally they cannot be described by a single mean-field order parameter. At zero temperature, this problem is not so severe, because one expects then both long-range order and on-site Bose-Fermi coherence, such that $\langle \hat{b} \rangle$ is large for both reasons. At finite temperature, however, this becomes a real problem, because the bosonic long range order is expected to get lost at temperatures on the order of the bosonic hopping t_b. The local Bose-Fermi coherence, on the other hand, remains intact for much higher temperatures, since the coupling g is orders of magnitudes higher. Indeed, we find that in the present approximation the full GDMFT calculations are in good agreement with a single site approximation. In this approach the impurity site does not any more couple neither to fermionic nor to the bosonic baths and long range order cannot be inferred. The critical temperature coming from this calculation, can be identified with the pair breaking temperature T_{pair}. This pair breaking temperature is very high and happens on a much higher energy scale than relevant for the experiment.

However, we want to address the question of long range order within this framework and calculate the critical temperature. In order to do so, we remark that the term $\Delta \hat{b}^\dagger$ in the Hamiltonian merely renormalizes the self-energy of the bosons: in the BEC regime the bosons are in a coherent state and this term is equivalent to a shift in the bosonic chemical potential. This is also clear from the treatment in [202], where the terms from the higher bands enter the bosonic self-energy. To make this more explicit, we write

$$\begin{aligned} \Delta &= -\sum_{l=1}^{\infty} g_l \langle \hat{c}_{l\uparrow} \hat{c}_{l\downarrow} \rangle = \pm \langle \hat{b} \rangle \left(\frac{g_0^2 \sqrt{\pi} \Gamma(-\bar{\mu}/\hbar\omega)/\Gamma(-\bar{\mu}/\hbar\omega - 1/2)}{\hbar\omega} \right. \\ &\left. + \sum_{l=0}^{N} \frac{g_l^2}{2(2l\hbar\omega - \bar{\mu})} - \sum_{l=1}^{N} \frac{g_l^2}{2\omega_l} \tanh\left(\frac{\omega_l}{2kT} \right) \right) \equiv \langle \hat{b} \rangle \Delta'. \end{aligned} \quad (6.32)$$

The above argument tells us that we can replace the term

$$-\left(\Delta \hat{b}^\dagger + h.c. \right) = -\left(\Delta' \langle \hat{b} \rangle \hat{b}^\dagger + h.c. \right) \quad (6.33)$$

in the Hamiltonian, by

$$-\Delta' \hat{b}^\dagger \hat{b}, \quad (6.34)$$

such that the term from the higher bands only renormalizes the chemical potential. We remark here, that this might look like an additional approximation. However, in view of the already made mean-field decoupling in the higher bands, this step in fact undoes part of the mean-field approximation. Alternatively this new form can be derived using second order perturbation theory in the higher bands couplings.

This alternative way to treat the higher bands gives for $T = 0$ almost identical results as before (see Fig. 6.2). The superfluid order parameter is a bit smaller, as expected. However, for nonzero temperatures this new scheme allows for a calculation of the critical temperature for long range order, which was not possible in the old scheme.

6.3 Results

We study a mixture of potassium atoms (^{40}K) and Feshbach molecules in a three-dimensional optical lattice. The on-site harmonic oscillator frequency is $\omega = 2\pi \times 58275$Hz, which corresponds to a lattice with wavelength $\lambda = 806$nm and Rabi frequency of $\Omega_R = 2\pi \times 1.43$GHz. The Feshbach resonance for Potassium is at $B = 202.1$G and the width of the resonance is 7.8 G. The difference of the magnetic moment between the closed and open channels of the Feshbach resonance is $\Delta\mu = 16/9\mu_B$, where μ_B is Bohr magneton. The total filling in our calculation is $n_{tot} = 1$.

First we consider zero temperature. Our calculations for the ground state are summarized in Fig. 6.2. Deep in the BEC regime only bosonic molecules are present. When the magnetic field is increased, close to resonance the number of fermions is increasing and the number of bosons decreasing. On the right from the Feshbach resonance we have mainly fermions and the number of bosons is small. The fermions are in the superconducting phase while the bosons are superfluid. In passing we remark that we describe the physics in terms of bare bosons and fermions here: in terms of dressed bosons, these are still bosons and the BEC/BCS crossover takes place when the bosonic self-energy crosses twice the Fermi energy [202]. However, in the case of half-filled fermions, this crossover is intercepted by a first order phase transition to a fermionic Mott insulator state. This happens at a critical value of the magnetic field of $B = 249$G. Calculations with only the lowest band of the Bose-Fermi-Hubbard model (as well as with one and two exited bands), showed this transition into the Mott insulator phase already close to the Feshbach resonance at $B \simeq 205$G. This implies that to capture the superconducting region 205T $\lesssim$ B < 249T the higher

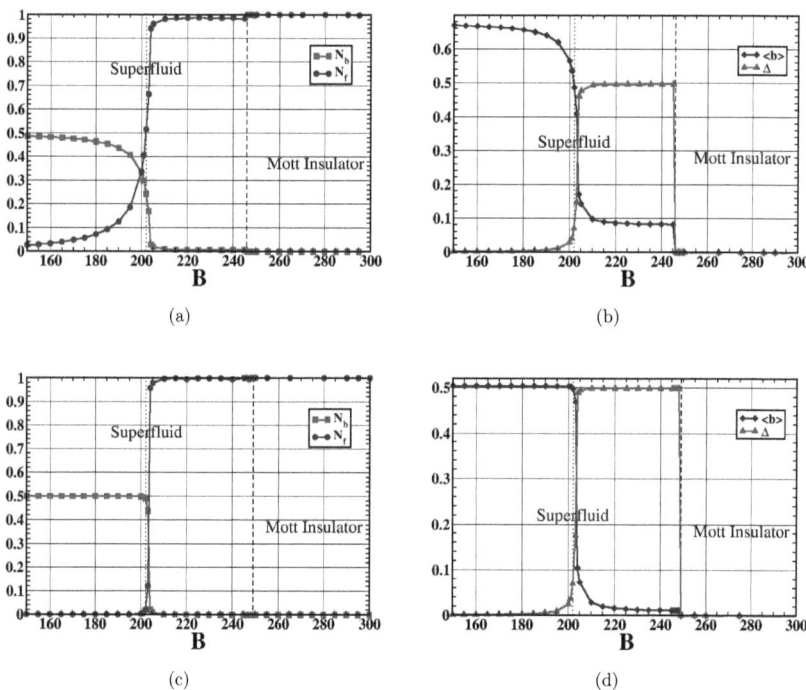

Figure 6.2: Here we plot the filling of fermions and bosons as well as superfluid and superconducting order parameters as a function of the magnetic field B for $T = 0$. In the sub-figure (a) and (b) we present results obtained by mean-field decoupling for both local and non-local correlations, while in the sub-figures (c) and (d) only non-local correlations are decoupled (for details see the text). In the sub-figures (a) and (c) we show the filling of fermions and bosons and in sub-figures (b) and (d) the superfluid and superconducting order parameters. Dotted line corresponds to the Feshbach resonance, while dashed line corresponds to the phase transition, from the superfluid/superconducting phase into the Mott insulator phase.

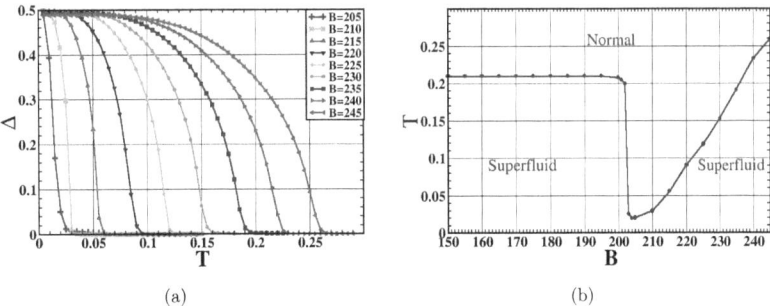

Figure 6.3: Finite temperature results: in subfigure (a) we plot the fermionic superfluid order parameter as a function of temperature T for different magnetic fields above the resonance, while in subfigure (b) we show the phase diagram. The blue solid line separates the superfluid phase from the normal phase. Here temperature is measured in units of the fermionic hopping t_f.

bands which renormalize the bosonic self-energy are crucial.

Having clarified the ground state phase diagram, we now consider finite temperature. In particular we investigate the critical temperature for the transition to the normal state. Deep in the BEC regime, the critical temperature is constant ($T_c \approx 0.21 t_f$) and completely determined by the properties of the bosons: the bosonic hopping parameter t_b, the interbosonic background scattering length a_b and the bosonic density. Only very close to resonance the critical temperature suddenly drops (see Fig. 6.3). This coincides with the magnetic field value for which fermions enter the system. On the BCS side of the resonance, the critical temperature depends on the magnetic field and increases with B (see Fig. 6.3). This implies that at resonance the critical temperature is *minimal*. This is in sharp contrast to the situation where no lattice is present, in which case the critical temperature is *maximal* close to resonance.

To summarize, the physics close to the Feshbach resonance is very rich. By changing the magnetic field, the ratio of fermionic and bosonic densities is changing. For lower magnetic field in the system are mainly bosons, close to the resonance the number of bosons is decreasing and the number of fermions increasing and the system contains mainly fermions. After further increase of the interaction transition to the Mott insulator phase takes place. At finite temperature we also observe a transition from the superfluid/superconducting phase to the normal phase. The physics seen here is even richer than in the case of the BEC-BCS crossover without the optical lattice, where the Mott Insulator phase is absent from the phase diagram.

Chapter 7

Summary

In this thesis we have studied the physics of different ultracold Bose-Fermi mixtures in optical lattices, as well as spin 1/2 fermions in a harmonic trap. To study these systems we generalized dynamical mean-field theory for a mixture of fermions and bosons, as well as for an inhomogeneous environment.

Generalized dynamical mean-field theory (GDMFT) is a method that describes a mixture of fermions and bosons. This method consists of Gutzwiller mean-field for the bosons, and dynamical mean-field theory for the fermions, which are coupled on-site by the Bose-Fermi density-density interaction and possibly a Feshbach term which converts a pair of up and down fermions into a molecule, i.e. a boson. We derived the self-consistency equations and showed that this method is well-controlled in the limit of high lattice coordination number z.

We develop real-space dynamical mean-field theory for studying systems in an inhomogeneous environment, e.g. in a harmonic trap. The crucial difference compared to standard DMFT is that we are taking into account that different sites are not equivalent to each other and thus take into account the inhomogeneity of the system. Different sites are coupled by the real-space Dyson equation.

We used these methods to study the following ultracold atomic systems:

Mixtures of fermions and bosons in optical lattices

First we considered a mixture of spinless fermions and bosons at zero temperature, when the filling of both of them is 1/2. We established the presence of a supersolid at weak Bose-Fermi repulsion. For strong interspecies interaction a first order phase transition occurs towards a state where the bosons are localized and form an alternating Mott insulator. An instability towards phase separation was observed for weak interaction among the bosons. As the gap in the spectral function for the supersolid phase is very small, we also studied the system for the case when fermions were half-filled, while the filling of the bosons was 3/2. We observed several phase transitions. A supersolid phase at small U_{fb} is succeeded by an alternating Mott insulator with alternating bosonic fillings 1 and 2 for larger U_{fb}. For even larger U_{fb} a second supersolid phase is stable, until for very large U_{fb} the ground state is formed by an AMI phase with alternating bosonic fillings 0 and 3. The quantum phase transitions found here are of second order, in contrast to the case of half-filled bosons, where a first-order quantum phase transition was observed. The phase diagram obtained in this case is particularly interesting because of the large amplitude of the supersolid density oscillations between the two AMI phases, which will make experimental observation easier.

We also investigated a mixture of two component fermions and hard-core bosons. For the case when interactions between fermions is stronger than interspecies interaction we obtain the antiferromagnetic phase, while in the other limit we obtained supersolid and AMI phases. The main finding in our calculations is the observation of the antiferromagnetic phase in the presence of the bosons. As the bosons can be used for cooling the system, this could facilitate the creation of an antiferromagnetic phase compared to a pure fermionic mixture.

Ultra-cold atoms in a harmonic trap

We used the R-DMFT to establish the stability of antiferromagnetism for balanced fermionic spin-$\frac{1}{2}$ systems in a trap and the appearance of phase separation for imbalanced mixtures.

We also study a mixture of spinless fermions and bosons in a harmonic trap. We have shown that the supersolid and AMI phases can be stabilized in an harmonic trap. However, the supersolid is very sensitive and its existence is easily destroyed by the trap, because of the small gap in the spectral function. The AMI phase, on the other hand, can be stabilized in the harmonic potential for a wide range of couplings.

Resonance superfluidity in an optical lattice

Finally we studied a system of fermions close to a s-wave Feshbach resonance. The physics close to the Feshbach resonance is very rich. By changing the magnetic field, the ratio of fermionic and bosonic numbers is varied. For lower magnetic field the system is mainly occupied by bosons, close to the resonance the number of bosons is decreasing and number of fermions increasing and the system contains mainly fermions. After further increase of the interaction a transition to the Mott insulator phase takes place if the fermions are at half-filling. At finite temperature we also observe a transition from the superfluid/superconducting phase to the normal phase. The physics seen here is even richer than in the case of the BEC-BCS crossover without the optical lattice, where the Mott Insulator phase is absent from the phase diagram.

Chapter 8

Zusammenfassung

Durch eindrucksvollen experimentellen Fortschritt innerhalb der letzten Dekade rückte das Gebiet der ultrakalten Atome in den Blickpunkt der Erforschung stark korrelierter Vielteilchensysteme. Die Fähigkeit Atome in optische Gitter, die durch stehende Wellen aus Laserlicht erzeugt werden, zu laden und darin zu manipulieren, macht es möglich, verschiedene Modell-Hamiltonoperatoren der Festkörperphysik zu realisieren. Dadurch können neue Einblicke in bekanntermaßen schwierigen Probleme [1–3] gewonnen werden. Darüberhinaus können Systeme realisiert werden, die keine Analoga in der Festkörperphysik besitzen. Im Besonderen können innerhalb des Gebietes der ultrakalten Atome Mischungen aus Bosonen und Fermionen erzeugt werden [4–19]. Diese Mischungen stellen physikalisch reichhaltige Systeme dar, die bisher noch wenig verstanden sind.

Bose-Fermi-Mischungen haben einige Ähnlichkeit zu den wohlbekannten zweikomponentigen Fermi-Fermi-Mischungen, jedoch besitzen sie eine reichhaltigere Physik. Indem man eine Spinkomponente durch ein Boson ersetzt, bewahrt man die Instabilität der Ladungsdichtewelle halbgefüllter fermionischer Systeme. Aus historischen Gründen wird diese Terminologie beibehalten, obwohl die betrachteten fermionischen Atome keine Ladung tragen. Zusätzlich können sich die Bosonen in der superfluiden Phase befinden, woraus sich die Möglichkeit supersoliden (SS) Verhaltens ergibt, welches gekennzeichnet ist durch koexistierende diagonale Ladungsdichtewellen und nichtdiagonale, langreichweitige Superfluidität. Zahlreiche theoretische Arbeiten haben Bose-Fermi-Mischungen in optischen Gittern als Forschungsgegenstand [20–45].

Zur Untersuchung stark korrelierter Bose-Fermi-Mischungen in optischen Gittern wurden verschiedene numerische und analytische Methoden benutzt. In einer Dimension beinhaltet dies den Bethe-Ansatz [25], Bosonisierung [26, 28], Dichtematrix-Renormierungsgruppe

[32, 35] und Quanten-Monte-Carlo [23, 36–40]. Jedoch sind nicht-perturbative Methoden in höheren Dimensionen rar. In zwei Dimensionen wurde die Renormierungsgruppenmethode verwendet [24, 31]. Obwohl diese Methode nicht-perturbative Effekte beschreiben kann, ist sie auf schwache Kopplungen beschränkt . Eine weitere Methode, das Ausintegrieren der Fermionen, wurde in zwei [29] und kürzlich in drei Dimensionen [33, 34] angewendet. Darin wird eine langreichweitige, retardierte Wechselwirkung zwischen den Bosonen erzeugt. Dies wiederum bedeutet, dass das resultierende bosonische Problem schwer zu lösen ist. Mit dieser Methode wurde ein wichtiger Fortschritt im Auffinden der Phasengrenzen zwischen Mott-Isolator und Superfluid erzielt. Weiterhin wurde ein Ansatz zusammengesetzter Fermionen verwendet, um qualitativ die möglichen Quantenphasen von Bose-Fermi-Mischungen zu beschreiben [30].

In dieser Arbeit wird die Physik ultrakalter Bose-Fermi-Mischungen in optischen Gittern und Spin-$\frac{1}{2}$-Fermionen in einer harmonischen Falle untersucht. Zu diesem Zweck wird die Dynamische-Molekularfeldtheorie für Bose-Fermi-Mischungen verallgemeinert, sowie hinsichtlich räumlich-inhomogener Umgebungen erweitert.

Die verallgemeinerte Dynamische-Molekularfeldtheorie (GDMFT) ist eine Methode zur Beschreibung von Bose-Fermi-Mischungen. Sie besteht aus dem Gutzwiller-Molekularfeld-Ansatz für die Bosonen und der Dynamischen-Molekularfeldtheorie für die Fermionen. Bosonen und Fermionen sind über eine Dichte-Dichte-Wechselwirkung am selben Gitterplatz gekoppelt. Betrachtet wird zudem auch eine Feshbach-Wechselwirkung, die ein Paar aus einem Spin-up und einem Spin-down Fermion in ein Molekül und damit in ein Boson umwandelt. Die Selbstkonsistenzgleichungen werden hergeleitet und es wird gezeigt, dass diese Methode eine über die Koordinationszahl des Gitters z gut kontrollierbare Näherung in hoher räumlicher Dimension darstellt. Da $1/z$ den einzig kleinen Parameter der Theorie darstellt, kann sie den ganzen Bereich von kleiner zu hoher Wechselwirkungsstärke beschreiben. Der Vorteil von GDMFT gegenüber QMC-Rechnungen ist der deutlich geringere Rechenaufwand in $d = 3$ Dimensionen. Die Methode erlaubt nicht nur die Phasengrenzen des Systems zu bestimmen, sondern darüberhinaus liefert sie auch verläßliche Ergebnisse abseits der Phasengrenzen (im Gegensatz zu den in [29, 33, 34] vorgestellten Methoden).

Die in dieser Arbeit entwickelte Ortsraum-Dynamische-Molekularfeldtheorie erlaubt es, Systeme in räumlich-inhomogener Umgebung, wie z.B. einer harmonischen Falle, zu untersuchen. Der wesentliche Unterschied zu Standard-DMFT ist die Berücksichtigung, dass die verschiedenen Gitterplätze nicht äquivalent sind. Die Gitterplätze sind über die

Ortsraum-Dyson-Gleichung gekoppelt.

Im Folgenden werden die ultrakalten atomaren Systeme vorgestellt, die mit den genannten Methoden untersucht werden.

Bose-Fermi-Mischungen in optischen Gittern

Zunächst werden Mischungen aus spinpolarisierten Fermionen und hard-core Bosonen am Temperaturnullpunkt untersucht. Das System ist jeweils halbgefüllt. Für kleine Interspezies-Wechselwirkung findet man eine supersolide Phase. Mit zunehmender Wechselwirkung geht das System via einen Phasenübergang erster Ordnung in eine Phase über, in der die Bosonen in einem alternierenden Mott-Isolator (AMI) lokalisiert sind. Darüberhinaus findet man eine Region in der beide Phasen koexistieren. Im die Energien der Zusände verglichen werden, kann festgestellt werden, dass die superfluide Phase den Grundzustand darstellt.

Danach wird die Annahme von vorliegenden hard-core Bosonen abgeschwächt und die repulsive Wechselwirkung zwischen den Bosonen als stark angenommen. Wiederum findet man eine supersolide Phase für schwache Interspezies-Wechselwirkung. Erhöht man diese, findet man eine Phasenübergang erster Ordnung in den alternierenden Mott-Isolator. Zudem wurde eine Instabilität gefunden, die zu Phasenseparation für schwache Wechselwirkung zwischen den Bosonen führt.

Die Anregungslücke in der fermionischen Spektralfunktion der supersoliden Phase ist sehr klein, wenn das bosonische System halbgefüllt ist. Somit ist diese Phase im Experiment schwierig zu detektieren. Um diese Schwierigkeit zu überwinden, wurde das System mit halber fermionischer Füllung und einer bosonischer Füllung von $n_b = 3/2$ studiert und verschiedene Phasenübergänge gefunden. Mit wachsender Interspezies-Wechselwirkung kommt man von einer supersoliden Phase zu einem Mott-Isolator mit alternierender bosonischer Füllung 1, bzw. 2, und dann zu einer zweiten supersoliden Phase gefolgt von einem alternierenden Mott-Isolator mit Füllung 0, bzw. 3. Es zeigt sich, dass es sich um Phasenübergänge zweiter Ordnung handelt im Gegensatz zu den Phasenübergängen erster Ordnung für das halbgefüllte bosonische System. Das errechnete Phasendiagramm ist besonders interessant, da die zweite supersolide Phase grosse Dichteoszillationen aufweist, die eine experimentelle Beobachtung dieser Phase erleichtert.

Desweiteren wird eine Mischung aus zweikomponentigen Fermionen und hard-core Bosonen untersucht. Wenn die Interspezies-Wechselwirkung viel schwächer ist als die Wechselwirkung zwischen den Fermionen erhält man Antiferromagnetismus (AF). Im umgekehrten Falle erhält man einen alternierenden Mott-Isolator. Im intermediären Regime findet man eine supersolide Phase und drei verschiedene Koexistenzbereiche: Koexistenz von (i) AF und SS, (ii) AF und AMI, sowie (iii) SS und AMI. Das Hauptergebnis der Untersuchung ist das Auffinden von Antiferromagnetismus in Gegenwart von Bosonen. Im Experiment können Bosonen genutzt werden, um das System zu kühlen. Dies könnte es, verglichen mit dem rein fermionischen system, erleichtern, Antiferromagnetismus in optischen Gittern experimentell zu beobachten.

Ultrakalte Atome in einer harmonischen Falle

Die experimentelle Beobachtung eines fermionischen Mott-Isolators in einer harmonischen Falle [89, 90] motiviert das Studium der Grundzustandseigenschaften eines solchen Systems. Es ist bekannt, dass das repulsive Hubbard-Model antiferromagnetische Ordnung bei kleinen Temperaturen aufweist. Experimentell kann eine solche Ordnung mittels Raman-Spektroskopie, durch Messen von Spinkorrelationsfunktionen, durch Messen von Korrelationen im Rauschen, Polarisationsspektroskopie oder Braggstreuung nachgewiesen werden. Mit Hilfe der R-DMFT werden in Zukunft Effekte in diesen Experimenten untersucht, die durch die räumliche Inhomogenität hervorgerufen werden. Darüberhinaus ermöglicht R-DMFT auch das Studium anderer starkkorrelierter System in räumlich-inhomogene Umgebungen.

R-DMFT wird verwendet, um die Stabilität des Antiferromagnetismus fermionischer Spin-$\frac{1}{2}$-Systeme in einer Falle nachzuweisen. Es wird gezeigt, dass eine antiferromagnetische Ordnung am Temperaturnullpunkt in einer Falle existiert. Die Ordnung ist stabil in Regionen, in denen die Gesamtfüllung nahezu eins ist. Sie besteht in Bereichen, in denen die Füllung signifikant von halber Füllung abweicht, fort. Je nach Parameterwahl kann die antiferromagnetische Ordnung im Zentrum der Falle oder als Ring, der eine paramagnetische Phase im Zentrum umgibt, realisiert werden.

Es werden die lokale Dichte und die lokale Magnetisierung innerhalb der Thomas-Fermi-Näherung (TFA) bestimmt, wobei ein externes Potential mit Hilfe eines variierenden chemischen Potentials [199] einbezogen wird. Eine gute Übereinstimmung zwischen den R-DMFT- und TFA-Ergebnissen wird in Bereichen klar innerhalb oder ausserhalb der antiferromagnetischen Domäne erzielt.

Es wird weiterhin gezeigt, dass für starke Repulsion Phasenseparation in unausgewogenen Mischungen auftritt, sofern der Unterschied in den Teilchenzahlen der beiden Spinkomponenten genügend groß ist. Für schwächere Repulsion verhindert eine starke Unausgewogenheit in den Teilchenzahlen eine antiferromagnetische Ordnung, jedoch führt sie nicht zu Phasenseparation. Diese Ergebnisse sind besonders faszinierend hinsichtlich neuerer Experimente mit attraktiv wechselwirkenden Fermionen, ein Thema zu dem noch einige unbeantwortete Fragen bezüglich der Natur der beobachteten Phasenseparation existieren.

Zudem wird eine Mischung aus spinpolarisierten Fermionen und Bosonen in einer harmonischen Falle untersucht. Es zeigt sich, dass die supersolide Phase und der alterniernde Mott-Isolator in einer harmonischen Falle stabilisiert werden können. Jedoch ist die supersolide Ordnung sehr schwach ausgeprägt, d.h. es liegt eine sehr kleine Anregungslücke vor, und kann leicht vom Einfluss der Falle zerstört werden, während der alternierende Mott-Isolator in einem großen Kopplungsbereich stabil ist.

Resonanz-Superfluidität in optischen Gittern

Abschliessend werden fermionische Systeme nahe der s-Wellen-Feshbach-Resonanz untersucht. Diese stellen ein anspruchsvolles Problem dar, da die Streulänge im Resonanzfall divergiert und somit der Hamilton-Operator in Gleichung (2.42) nicht definiert ist. Dieses Problem löst man, indem man einen Hamilton-Opertor definiert, in dem der Resonanzfall separiert und explizit behandelt wird [200]. Die nicht-resonanten Anteile führen zu einer Hintergrundstreulänge und somit zu Wechselwirkungen zwischen den Fermionen. Da die Feshbach-Resonanz aufgrund einer Kopplung mit einem bosonischen Molekülzustand auftritt, muss man explizit bosonische Freiheitsgrade in der Theorie einbinden [200], um resonante Prozesse zu beschreiben .

Ein ultrakaltes atomares Gas aus fermionischen Atomen und bosonischen Molekülen in einem optischen Gitter nahe der Feshbach-Resonanz kann über das Bose-Fermi-Hubbard-Modell [201, 202] beschrieben werden. In der vorliegenden Arbeit wird angenommen, dass alle Bosonen ausschließlich Zustände im untersten Bond besetzen. Hinsichtlich der Fermionen andererseits müssen auch höhere Bänder beachtet werden, da die Feshbach-Physik in einem optischen Gitter von Natur aus ein Multiband-Problem [202, 203] darstellt. Da die Bandbreite viel kleiner als die Bandlücke ist, können die höheren Bänder als flach angenommen werden und die volle Bandstruktur muss nur für das unterste Band beachtet werden. Darüberhinaus werden Wechselwirkungen zwischen Fermionen in höheren Bändern, sowie deren Wechselwirkung mit Bosonen vernachlässigt. Diese Annahme ist gerechtfertigt, da die Füllung in höheren Bändern sehr gering ist, so dass Wechsewirkungseffekte klein sind. Der Hamilton-Operator des Systems ist somit durch die in Gleichung (6.1) aufgeführte Form gegeben.

Zur Untersuchung des Systems wird GDMFT verwendet. Indem man das magnetische Feld variiert, ändert man das Verhältnis aus Fermionenanzahl zu der Bosonenanzahl. Für kleines magnetisches Feld besteht das System hauptsächlich aus Bosonen. Nahe der Resonanz nimmt die Bosonenanzahl ab, während die Fermionenanzahl zunimmt, bis das System schließlich hauptsächlich aus Fermionen besteht. Weiteres Erhöhen der Wechselwirkungsstärke führt zu einem Übergang zu einem Mott-Isolator, falls das fermionische System halbgefüllt ist. Für endliche Temperaturen erhält man einen Übergang zwischen superfluider und normaler Phase. Das untersuchte System weist eine vielfältigere Physik auf, als der BEC-BCS-Übergang ohne optisches Gitter, da dort kein Mott-Isolator auftritt.

Appendix A

Relation between Experimental and Hubbard Parameters

In this Appendix we derive the relations between the experimental and Hubbard parameters. The Hamiltonian for a Bose-Fermi mixture in second quantized form is given as

$$\hat{\mathcal{H}} = \hat{T}_f + \hat{T}_b + \hat{V}_f + \hat{V}_b + \hat{W}_{ff} + \hat{W}_{bb} + \hat{W}_{fb}, \qquad (A.1)$$

where

$$\hat{T}_f = -\sum_\sigma \int d^3\mathbf{r}\, \hat{\Psi}^\dagger_{f\sigma}(\mathbf{r}) \frac{\hbar^2 \nabla^2}{2m_f} \hat{\Psi}_{f\sigma}(\mathbf{r}), \qquad (A.2)$$

$$\hat{T}_b = -\int d^3\mathbf{r}\, \hat{\Psi}^\dagger_b(\mathbf{r}) \frac{\hbar^2 \nabla^2}{2m_b} \hat{\Psi}_b(\mathbf{r}), \qquad (A.3)$$

$$\hat{W}_{ff} = \int \hat{\Psi}^\dagger_{f\downarrow}(\mathbf{r}) \hat{\Psi}^\dagger_{f\uparrow}(\mathbf{r}) \frac{4\pi\hbar^2 a_f}{m_f} \hat{\Psi}_{f\uparrow}(\mathbf{r}) \hat{\Psi}_{f\downarrow}(\mathbf{r}), \qquad (A.4)$$

$$\hat{W}_{bb} = \frac{1}{2}\int \hat{\Psi}^\dagger_b(\mathbf{r}) \hat{\Psi}^\dagger_b(\mathbf{r}) \frac{4\pi\hbar^2 a_b}{m_b} \hat{\Psi}_b(\mathbf{r}) \hat{\Psi}_b(\mathbf{r}), \qquad (A.5)$$

$$\hat{W}_{fb} = \sum_\sigma \int \hat{\Psi}^\dagger_{f\sigma}(\mathbf{r}) \hat{\Psi}^\dagger_b(\mathbf{r}) \frac{2\pi\hbar^2 a_{fb}}{m_r} \hat{\Psi}_b(\mathbf{r}) \hat{\Psi}_{f\sigma}(\mathbf{r}), \qquad (A.6)$$

$$\hat{V}_f = \sum_\sigma \int d^3\mathbf{r}\, \hat{\Psi}^\dagger_{f\sigma}(\mathbf{r}) V_f(\mathbf{r}) \hat{\Psi}_{f\sigma}(\mathbf{r}), \qquad (A.7)$$

$$\hat{V}_b = \int d^3\mathbf{r}\, \hat{\Psi}^\dagger_b(\mathbf{r}) V_b(\mathbf{r}) \hat{\Psi}_b(\mathbf{r}). \qquad (A.8)$$

Here $\hat{\Psi}^\dagger_{f\sigma}(\mathbf{r})$ is the creation operator of a fermion with spin σ at point $\mathbf{r}$, while $\hat{\Psi}^\dagger_b(\mathbf{r})$ describes the creation operator of a boson at point $\mathbf{r}$. a_f, a_b and a_{fb} are the s-wave scattering lengths for Fermi-Fermi, Bose-Bose and Bose-Fermi interactions, respectively. m_f (m_b) is the mass of fermions (bosons) and $m_r = m_f m_b/(m_f + m_b)$. V_f and V_b denotes the external potential for fermions and bosons, respectively.

In the presence of the periodic potential $V_{f(b)}(\mathbf{r} + \mathbf{R}) = V_{f(b)}(\mathbf{r})$, it is convenient to express the fermionic (bosonic) creation operators $\hat{\Psi}^\dagger_{f\sigma}(\mathbf{r})$ ($\hat{\Psi}^\dagger_b(\mathbf{r})$) using Wannier functions:

$$\hat{\Psi}^\dagger_{f\sigma} = \sum_{i,l} \hat{c}^\dagger_{i\sigma l} w^f_{l,x}(x - x_i) w^f_{l,y}(y - y_i) w^f_{l,z}(z - z_i), \tag{A.9}$$

$$\hat{\Psi}^\dagger_b = \sum_{i,l} \hat{b}^\dagger_{i,l} w^b_{l,x}(x - x_i) w^b_{l,y}(y - y_i) w^b_{l,z}(z - z_i), \tag{A.10}$$

where $\hat{c}^\dagger_{i\sigma}$ ($\hat{b}^\dagger_{i,l}$) are fermionic (bosonic) creation operator at site $\mathbf{r}_i = (x_i, y_i, z_i)$ and $w^{f(b)}_l(\mathbf{r} - \mathbf{r}_i) = w^{f(b)}_{l,x}(x - x_i) w^{f(b)}_{l,y}(y - y_i) w^{f(b)}_{l,z}(z - z_i)$ is the Wannier function for a localized fermion (boson) in the lth energy band.

For temperatures and interactions much smaller than the band gap, only the lowest band will be occupied. This means that the sum over the band indices l is reduced to $l = 0$ and we can drop the band index.

If one puts Eq. (A.9) and (A.10) into equations (A.2-A.8) one will obtain that:

$$\hat{\mathcal{H}} = -t_f \sum_{\langle ij \rangle, \sigma} \hat{c}^\dagger_{i\sigma} \hat{c}_{j\sigma} - t_b \sum_{\langle ij \rangle} \hat{b}^\dagger_i \hat{b}_j + U_f \sum_i \hat{n}^f_{i\uparrow} \hat{n}^f_{i\downarrow} + \frac{U_b}{2} \sum_i \hat{n}^b_i(\hat{n}^b_i - 1) + U_{fb} \sum_i \hat{n}^f_i \hat{n}^b_i, \tag{A.11}$$

where

$$t_f = \int d^3 r\, w^f_x(x - x_i + a) w^f_y(y - y_i) w^f_z(z - z_i) \frac{\hbar^2 \nabla^2}{2m_f} w^f_x(x - x_i) w^f_y(y - y_i) w^f_z(z - z_i), \tag{A.12}$$

$$t_b = \int d^3 r\, w^b_x(x - x_i) w^b_y(y - y_i) w^b_z(z - z_i) \frac{\hbar^2 \nabla^2}{2m_b} w^b_x(x - x_i + a) w^b_y(y - y_i) w^b_z(z - z_i), \tag{A.13}$$

$$U_f = \frac{4\pi \hbar^2 a_f}{m_f} \int dx |w^f_x(x - x_i)|^4 \int dy |w^f_y(y - y_i)|^4 \int dz |w^f_z(z - z_i)|^4, \tag{A.14}$$

$$U_b = \frac{4\pi \hbar^2 a_b}{m_b} \int dx |w^b_x(x - x_i)|^4 \int dy |w^b_y(y - y_i)|^4 \int dz |w^b_z(z - z_i)|^4, \tag{A.15}$$

$$U_{fb} = \frac{2\pi \hbar^2 a_{bf}}{m_r} \int dx \left[w^f_x(x - x_i) w^b_x(x - x_i) \right]^2 \int dy \left[w^f_y(y - y_i) w^b_y(y - y_i) \right]^2$$
$$\times \int dz \left[w^f_z(z - z_i) w^b_z(z - z_i) \right]^2. \tag{A.16}$$

For a deep optical lattices one can approximate the Wannier functions by Gaussian, i.e:

$$w_\alpha^{f(b)}(\alpha) = \frac{e^{-\alpha^2/2l_{f(b)}^2}}{\pi^{1/4} l_{f(b)}^{1/2}}, \qquad (A.17)$$

where $\alpha = x, y, z$, and

$$l_{f(b)} = \frac{a}{\pi (V_0^{f(b)}/E_r^{f(b)})^{1/4}}, \qquad (A.18)$$

where $E_r^{f(b)} = \hbar^2/2\lambda^2 m_{f(b)}$ is a recoil energy, $V_0^{f(b)}$ is the laser potential strength and λ is the wavelength of laser.

We will not calculate here the hopping coefficients, because for this purpose we have to go beyond a Gaussian Anstatz, since for the hopping coefficient the overlap with the neighboring wave-function is important and the Gaussian function only approximates the Wannier function well close to the potential minimum. The correct expression can be obtained from the asymptotically exact solution of the 1D Mathieu equation for a deep lattice [143]:

$$t_{f(b)} = \frac{4}{\sqrt{\pi}} E_r^{f(b)} \left(\frac{V_0^{f(b)}}{E_r^{f(b)}}\right)^{3/4} \exp\left[-2\left(\frac{V_0^{f(b)}}{E_r^{f(b)}}\right)\right]. \qquad (A.19)$$

Now we calculate U_f and U_b. Using Eqs. (A.14), (A.15) and (A.17), (A.18) we obtain

$$\begin{aligned}
U_{f(b)} &= \frac{4\pi\hbar^2 a_{f(b)}}{m_{f(b)}} \left[\int dx \left|w_x^{f(b)}(x)\right|^4\right]^3 = \frac{4\pi\hbar^2 a_{f(b)}}{m_{f(b)}} \left(\frac{1}{\pi l_{f(b)}^2}\right)^3 \left[\int dx\, e^{-\frac{2}{l_{f(b)}^2} x^2}\right]^3 \\
&= \frac{4\pi\hbar^2 a_{f(b)}}{m_{f(b)}} \left(\frac{1}{\pi l_{f(b)}^2}\right)^3 \left(\frac{\pi l_{f(b)}^2}{2}\right)^{3/2} = \frac{4\pi\hbar^2 a_b}{m_{f(b)}} \frac{1}{\pi\sqrt{8\pi} l_{f(b)}^3} = \frac{4\hbar^2 a_{f(b)}}{m_b \sqrt{8\pi}} \frac{\pi^3 (V_0^{f(b)}/E_r^{f(b)})^{3/4}}{a^3} \\
&= \sqrt{\frac{8}{\pi}} k a_{f(b)} E_r^{f(b)} \left(\frac{V_0^{f(b)}}{E_r^{f(b)}}\right)^{3/4}. \qquad (A.20)
\end{aligned}$$

Finally we calculate U_{fb}. Using Eqs. (A.16), (A.17) and (A.18) leads to

$$\begin{aligned}
U_{fb} &= \frac{2\pi\hbar^2 a_{bf}}{m_r} \left[\int dx \left[w_x^f(x) w_x^b(x) \right]^2 \right]^3 = \frac{2\pi\hbar^2 a_{bf}}{m_r} \left(\frac{1}{\pi l_f l_b} \right)^3 \left[\int dx e^{-\frac{l_f^2 + l_b^2}{l_f^2 l_b^2} x^2} \right]^3 \\
&= \frac{2\pi\hbar^2 a_{bf}}{m_r} \left(\frac{1}{\pi l_f l_b} \right)^3 \left[\pi \frac{l_f^2 l_b^2}{l_f^2 + l_b^2} \right]^{3/2} = \frac{2\hbar^2 a_{bf}}{m_r \sqrt{\pi}} \frac{1}{\left(l_f^2 + l_b^2 \right)^{3/2}} \\
&= \frac{2\hbar^2 a_{bf}}{m_r \sqrt{\pi}} \frac{\pi^3}{a^3} \frac{1}{\left((E_r^f/V_0^f)^{1/2} + (E_r^b/V_0^b)^{1/2} \right)^{3/2}} = \frac{4k a_{fb} E_r^b}{\sqrt{\pi}} \left(1 + \frac{m_b}{m_f} \right) \frac{1}{\left(1 + \frac{(E_r^f V_0^b)^{1/2}}{(E_r^b V_0^f)^{1/2}} \right)^{3/2}} \left(\frac{V_0^b}{E_r^b} \right)^{3/4} \\
&= \frac{4k a_{fb} E_r^b}{\sqrt{\pi}} \frac{1 + m_b/m_f}{\left(1 + \sqrt{m_b V_0^b / m_f V_0^f} \right)^{3/2}} \left(\frac{V_0^b}{E_r^b} \right)^{3/4}. \quad\quad\quad (A.21)
\end{aligned}$$

Appendix B

Derivation of the DMFT Effective Action

To derive the DMFT self-consistency relations for the Hamiltonian (3.23), we use the path integral formalism. The partition function is given by :

$$Z = \int \prod_{i,\sigma} D\tilde{c}^{\star}_{i\sigma} D\tilde{c}_{i\sigma} D\tilde{b}^{\star}_i D\tilde{b}_i \, e^{-S}. \tag{B.1}$$

The action is written as $S = S_0 + S^o + \Delta S$, with

$$\begin{aligned}
S_0 &= \int_0^\beta d\tau \Big\{ \sum_\sigma \tilde{c}^{\star}_{0\sigma}(\partial_\tau - \mu_{\sigma f})\tilde{c}_{0\sigma} + \tilde{b}^{\star}_0(\partial_\tau - \mu_b)\tilde{b}_0 \\
&\quad + U_f \tilde{n}^f_{0\uparrow}\tilde{n}^f_{0\downarrow} + \frac{U_b}{2}\tilde{n}^b_0(\tilde{n}^b_0 - 1) + U_{fb}n^f_0 n^b_0 + g(\tilde{c}^{\star}_{0\uparrow}\tilde{c}^{\star}_{0\downarrow}\tilde{b}_0 + c.c) \Big\}, \\
\Delta S &= -\int_0^\beta d\tau \Big\{ t_f \sum_{i\sigma}{}' (\tilde{c}^{\star}_{0\sigma}\tilde{c}_{i\sigma} + \tilde{c}^{\star}_{i\sigma}\tilde{c}_{0\sigma}) \\
&\quad + t_b \sum_i{}' (\tilde{b}^{\star}_0 \tilde{b}_i + \tilde{b}^{\star}_i \tilde{b}_0) \Big\}, \\
S^o &= \int_0^\beta d\tau \Big\{ -t_f \sum_{\langle ij \rangle^o \sigma} \tilde{c}^{\star}_{i\sigma}\tilde{c}_{j\sigma} - t_b \sum_{\langle ij \rangle^o} \tilde{b}^{\star}_i \tilde{b}_j \\
&\quad + \sum_{i \neq 0} \Big(U_f \tilde{n}^f_{i\uparrow}\tilde{n}^f_{i\downarrow} + \frac{U_b}{2}\tilde{n}^b_i(\tilde{n}^b_i - 1) + U_{fb}\tilde{n}^f_i \tilde{n}^b_i + g(\tilde{c}^{\star}_{i\downarrow}\tilde{c}^{\star}_{i\uparrow}\tilde{b}_i + c.c) \Big) \Big\},
\end{aligned} \tag{B.2}$$

where β is the inverse temperature, τ is imaginary time, $\tilde{c}_{i\sigma}$, $\tilde{c}^{\star}_{i\sigma}$ are Grassmann variables describing the fermions, $\tilde{b}_i$, $\tilde{b}^{\star}_i$, $\tilde{n}^b_i$, $\tilde{n}^f_i$ are $\mathbb{C}$-numbers describing the bosons and the number of fermions/bosons. Here the action is divided into three parts. S_0 describes the "impurity site", S^o describes the system without the impurity and ΔS is the coupling between them. $\sum'$ means that the summations run only over the nearest neighbors of the "impurity site" and $\langle ij \rangle^o$ indicates a summation over all pairs of nearest neighbor sites excluding the impurity site (i.e. $i, j \neq 0$).

We now derive an effective action for the impurity, defined by

$$\frac{1}{Z_{eff}}e^{-S_{eff}} \equiv \frac{1}{Z}\int \prod_{i \neq 0, \sigma} D\tilde{c}^*_{i\sigma}D\tilde{c}_{i\sigma}D\tilde{b}^*_i D\tilde{b}_i\, e^{-S}. \tag{B.3}$$

Using Eqs. (B.1), (B.2) and (B.3) and with the definition $\Delta S = \int d\tau \Delta S(\tau)$ we obtain

$$\frac{e^{-S_{eff}}}{Z_{eff}} = \frac{e^{-S_0}}{Z}\int \prod_{i \neq 0, \sigma} D\tilde{c}^*_{i\sigma}D\tilde{c}_{i\sigma}D\tilde{b}^*_i D\tilde{b}_i\, e^{-S^o}e^{-\Delta S} = \frac{e^{-S_0}}{Z}\int \prod_{i \neq 0, \sigma} D\tilde{c}^*_{i\sigma}D\tilde{c}_{i\sigma}D\tilde{b}^*_i D\tilde{b}_i\, e^{-S^o} \sum_{n=0}^{\infty} \frac{(-\Delta S)^n}{n!}$$

$$= e^{-S_0}\frac{Z^o}{Z}\left(1 - \int_0^{\beta} d\tau \langle \Delta S(\tau)\rangle^o + \frac{1}{2!}\int_0^{\beta} d\tau_1 \int_0^{\beta} d\tau_2 \langle \Delta S(\tau_1)\Delta S(\tau_2)\rangle^o + \ldots \right)$$

$$= e^{-S_0}\frac{Z^o}{Z}\left(1 + t_b\int_0^{\beta} d\tau \sum_i {}'(\Phi_i^o(\tau)\tilde{b}_0^\star(\tau) + c.c) - t_f^2 \int_0^{\beta} d\tau_1 \int_0^{\beta} d\tau_2 \sum_{i,j,\sigma}{}' \tilde{c}^*_{0\sigma}(\tau_1)G^o_{ij,\sigma}(\tau_1 - \tau_2)\tilde{c}_{0\sigma}(\tau_2)\right.$$

$$\left. -\frac{1}{2}t_b^2 \int_0^{\beta} d\tau_1 \int_0^{\beta} d\tau_2 \sum_{i,j}{}' \tilde{\mathbf{b}}_0^\star(\tau_1)\mathbf{G}^o_{b,ij}(\tau_1 - \tau_2)\tilde{\mathbf{b}}_0(\tau_2) + \ldots \right), \tag{B.4}$$

where Z^o is the statistical sum without the "impurity" site and $\langle \ldots \rangle^o$ are expectation values in the system not including the "impurity site". We have introduced the Nambu-space vector $\tilde{\mathbf{b}}_0(\tau) = \begin{pmatrix} \tilde{b}_0(\tau) \\ \tilde{b}_0^\star(\tau) \end{pmatrix}$, $\Phi_i^o(\tau) = \langle \hat{b}_i(\tau)\rangle^o$ as the bosonic superfluid parameter, $G^o_{ij,\sigma}(\tau_1 - \tau_2) = -\langle T\hat{c}_{i\sigma}(\tau_1)\hat{c}^\dagger_{j\sigma}(\tau_2)\rangle^o$ as the Green's function for the fermions and $\mathbf{G}^o_{b,ij}(\tau_1 - \tau_2) = -\left\langle T\begin{pmatrix} \hat{b}_i(\tau_1) \\ \hat{b}^\dagger_i(\tau_1) \end{pmatrix} \left(\hat{b}^\dagger_j(\tau_2), \hat{b}_j(\tau_2)\right)\right\rangle^o$ as the Green's function for the bosons in Nambu space.

By the linked-cluster theorem we obtain

$$S_{eff} = S_0 - t_b\int_0^{\beta} d\tau \sum_i{}' (\Phi_i^o(\tau)\tilde{b}_0^\star(\tau) + c.c) + t_f^2 \sum_{\sigma}\int_0^{\beta} d\tau_1 \int_0^{\beta} d\tau_2 \sum_{i,j}{}' \tilde{c}^*_{0\sigma}(\tau_1)G^o_{ij,\sigma}(\tau_1 - \tau_2)\tilde{c}_{0\sigma}(\tau_2)$$

$$+\frac{1}{2}t_b^2 \int_0^{\beta} d\tau_1 \int_0^{\beta} d\tau_2 \sum_{i,j}{}' \tilde{\mathbf{b}}_0^\star(\tau_1)\mathbf{G}^o_{b,ij}(\tau_1 - \tau_2)\tilde{\mathbf{b}}_0(\tau_2) + \ldots . \tag{B.5}$$

In this sum also higher order correlation functions appear (indicated by the dots).

In order to retain a finite kinetic energy, the hopping parameters should be rescaled.

The bosonic hopping parameter should be rescaled as $t_b \to t_b = t_b^*/z$, and only the leading bosonic term describing the coupling to the bosonic superfluid order parameter survives in infinite dimensions. The fermionic hopping parameter will be rescaled as $t_f \to t_f = t_f^*/\sqrt{z}$ according to fermionic DMFT. [147, 148]. After rescaling the hopping parameters and considering the limit $z \to \infty$ only the leading term for fermions and bosons survives. We obtain that Eq. (B.5) reduces to the following relation:

$$S_{eff} = S_0 - t_b \int_0^\beta d\tau {\sum_i}' (\Phi_i^o(\tau)\tilde{b}_0^\star(\tau) + c.c) \quad \text{(B.6)}$$
$$+ t_f^2 \sum_\sigma \int_0^\beta d\tau_1 \int_0^\beta d\tau_2 {\sum_{ij}}' \tilde{c}_{0\sigma}^\star(\tau_1) G_{ij,\sigma}^o(\tau_1 - \tau_2) \tilde{c}_{0\sigma}(\tau_2).$$

Appendix C

The Equation of Motion and Green's Functions

Here we derive the equation of motion for the Green's functions [206]. We consider two arbitrary operators $\hat{A}$ and $\hat{B}$. These operators are either fermionic or bosonic. We work in the Matsubara frequency representation. In this representation the Greens' function is defined as follows:

$$G_{\hat{A}\hat{B}}(\tau) \equiv \langle\langle \hat{A}, \hat{B} \rangle\rangle_\tau = -\langle T_\tau \hat{A}(\tau) \hat{B}(0) \rangle = -\theta(\tau)\langle \hat{A}(\tau)\hat{B}(0)\rangle + \eta\theta(-\tau)\langle \hat{B}(0)\hat{A}(\tau)\rangle, \quad (C.1)$$

where $\eta = +1$ for fermions and $\eta = -1$ for bosons.

From Eq. (C.1) we directly obtain that

$$\frac{d}{d\tau}\hat{G}_{\hat{A}\hat{B}}(\tau) = -\delta(\tau)\langle[\hat{A},\hat{B}]_\eta\rangle - \langle T_\tau \frac{d\hat{A}(\tau)}{d\tau}\hat{B}(0)\rangle = -\delta(\tau)\langle[\hat{A},\hat{B}]_\eta\rangle + G_{[\hat{H},\hat{A}],\hat{B}}(\tau). \quad (C.2)$$

Here we used the relation for the derivative of an operator which does not explicitly depend on the time :

$$\frac{d\hat{A}(\tau)}{d\tau} = [\hat{\mathcal{H}}, \hat{A}]_- . \quad (C.3)$$

Eq. (C.2) is the equation of motion in the imaginary time representation. Now we perform the Fourier Transformation :

$$G_{\hat{A},\hat{B}}(\tau) = \frac{1}{\beta}\sum_{n=-\infty}^{\infty} e^{-i\omega_n \tau} G_{\hat{A},\hat{B}}(i\omega_n), \quad (C.4)$$

with the Matsubara frequencies $\omega_n = \frac{(2n+1)\pi}{\beta}$ for the fermions and $\omega_n = \frac{2n\pi}{\beta}$ for the bosons, and rewrite the equation of motion in the Matsubara frequency representation:

$$i\omega_n G_{\hat{A},\hat{B}}(i\omega_n) + G_{[\hat{\mathcal{H}},\hat{A}]_-,\hat{B}}(i\omega_n) = \langle [\hat{A},\hat{B}]_\eta \rangle, \qquad (C.5)$$

As one can see, the equation of motion Eq. (C.5) maps the initial Green's function to new Green's function(s). For these new Green's function(s) we have to again use the equation of motion, which produces again new Green's function(s), but as we see in the next appendix, this circle is closed and we are able to determine all of these Green's functions.

In the end of this appendix we present several relations for Green's functions which we use later. From Eq. (C.1) one can easily derive that:

$$G_{\hat{A},\hat{B}}(\tau) = -\eta G_{\hat{B},\hat{A}}(-\tau), \qquad (C.6)$$

$$G_{\hat{A},\hat{B}}(\tau \pm \beta) = -\eta G_{\hat{A},\hat{B}}(\tau) = G_{\hat{B},\hat{A}}(-\tau), \qquad (C.7)$$

$$\left(G_{\hat{A},\hat{B}}(\tau)\right)^* = -\eta G_{\hat{A}^\dagger,\hat{B}^\dagger}(-\tau) = G_{\hat{B}^\dagger,\hat{A}^\dagger}(\tau). \qquad (C.8)$$

Applying the inverse Fourier transformation

$$G_{\hat{A},\hat{B}}(i\omega_n) = \int_0^\beta d\tau e^{i\omega_n \tau} G_{\hat{A},\hat{B}}(\tau) \qquad (C.9)$$

in Eqs. (C.6), (C.7), and (C.8) leads to

$$G_{\hat{A},\hat{B}}(i\omega_n) = -\eta G_{\hat{B},\hat{A}}(-i\omega_n), \qquad (C.10)$$

$$\left(G_{\hat{A},\hat{B}}(i\omega_n)\right)^* = -\eta G_{\hat{A}^\dagger,\hat{B}^\dagger}(i\omega_n) = G_{\hat{B}^\dagger,\hat{A}^\dagger}(-i\omega_n). \qquad (C.11)$$

Appendix D

Derivation of the Self-Energy for a Bose-Fermi Mixture

In this appendix we evaluate the self-energy via correlation functions. For this purpose we are using the equation of motion, which has the following form:

$$i\omega_n \langle\langle \hat{A}, \hat{B} \rangle\rangle_\omega + \langle\langle \left[\hat{\mathcal{H}}, \hat{A}\right]_-, \hat{B} \rangle\rangle_\omega = \langle \left[\hat{A}, \hat{B}\right]_\eta \rangle, \tag{D.1}$$

where ω_n are the Matsubara frequencies and $\langle\langle \hat{A}, \hat{B} \rangle\rangle_\omega$ is the general Green's function.

For the Fermi-Bose mixture a generalized single impurity Anderson Hamiltonian has the following form:

$$\hat{\mathcal{H}} = -\sum_\sigma \mu_{f\sigma} \hat{n}^f_\sigma + U_f \hat{n}^f_\uparrow \hat{n}^f_\downarrow + U_{fb} \hat{n}^f \hat{n}^b + g\left(\hat{f}^\dagger_\downarrow \hat{f}^\dagger_\uparrow \hat{b} + h.c.\right) + \hat{\mathcal{H}}_B$$
$$+ \sum_{k\sigma} V_{k\sigma} \left(\hat{f}^\dagger_\sigma \hat{c}_{k\sigma} + h.c\right) + \sum_{k\sigma} \varepsilon_{k\sigma} \hat{c}^\dagger_{k\sigma} \hat{c}_{k\sigma} + \sum_k W_k \left(\hat{c}^\dagger_{k\uparrow} \hat{c}^\dagger_{k\downarrow} + h.c\right), \tag{D.2}$$

where $\hat{f}^\dagger_\sigma$ and $\hat{c}^\dagger_{k\sigma}$ are the fermionic creation operators on the "impurity cite" and "conduction band" respectively. $\hat{b}^\dagger$ is the bosonic creation operator on the impurity cite. $\hat{n}^f = \hat{n}^f_\uparrow + \hat{n}^f_\downarrow = \sum_\sigma \hat{f}^\dagger_\sigma \hat{f}_\sigma$, $\hat{n}^b = \hat{b}^\dagger \hat{b}$ and $\hat{\mathcal{H}}_B$ is bosonic part of the Hamiltonian.

To solve the problem we should first calculate the following commutation relations

$$\left[\hat{\mathcal{H}}, \hat{f}_\sigma\right]_- = \mu_{f\sigma}\hat{f}_\sigma - U_f \hat{f}_\sigma \hat{f}_{\bar{\sigma}}^\dagger \hat{f}_{\bar{\sigma}} - U_{fb}\hat{f}_\sigma \hat{b}^\dagger \hat{b} + \sigma g \hat{f}_{\bar{\sigma}}^\dagger \hat{b} - \sum_k V_{k\sigma}\hat{c}_{k\sigma}, \tag{D.3}$$

$$\left[\hat{\mathcal{H}}, \hat{f}_\sigma^\dagger\right]_- = -\mu_{f\sigma}\hat{f}_\sigma^\dagger + U_f \hat{f}_\sigma^\dagger \hat{f}_{\bar{\sigma}}^\dagger \hat{f}_{\bar{\sigma}} + U_{fb}\hat{f}_\sigma^\dagger \hat{b}^\dagger \hat{b} - \sigma g \hat{f}_{\bar{\sigma}} \hat{b}^\dagger + \sum_k V_{k\sigma}\hat{c}_{k\sigma}^\dagger, \tag{D.4}$$

$$\left[\hat{\mathcal{H}}, \hat{c}_{k\sigma}\right]_- = -\varepsilon_{k\sigma}\hat{c}_{k\sigma} - V_{k\sigma}\hat{f}_\sigma - \sigma W_k \hat{c}_{k\bar{\sigma}}^\dagger, \tag{D.5}$$

$$\left[\hat{\mathcal{H}}, \hat{c}_{k\sigma}^\dagger\right]_- = \varepsilon_{k\sigma}\hat{c}_{k\sigma}^\dagger + V_{k\sigma}\hat{f}_\sigma^\dagger + \sigma W_k \hat{c}_{k\bar{\sigma}}, \tag{D.6}$$

where $\bar{\sigma} = -\sigma$.

First we use the equation of motion for the case when $\hat{A} = \hat{f}_\sigma$ and $\hat{B} = \hat{f}_\sigma^\dagger$. Putting commutation relation Eq. (D.3) in the equation of motion Eq. (D.1) we get:

$$(i\omega_n + \mu_{f\sigma})\langle\langle \hat{f}_\sigma, \hat{f}_\sigma^\dagger \rangle\rangle_\omega - U_f \langle\langle \hat{f}_\sigma \hat{f}_{\bar{\sigma}}^\dagger \hat{f}_{\bar{\sigma}}, \hat{f}_\sigma^\dagger \rangle\rangle_\omega - U_{fb}\langle\langle \hat{f}_\sigma \hat{b}^\dagger \hat{b}, \hat{f}_\sigma^\dagger \rangle\rangle_\omega$$
$$+ \sigma g \langle\langle \hat{f}_{\bar{\sigma}}^\dagger \hat{b}, \hat{f}_\sigma^\dagger \rangle\rangle_\omega - \sum_k V_{k\sigma}\langle\langle \hat{c}_{k\sigma}, \hat{f}_\sigma^\dagger \rangle\rangle_\omega = 1. \tag{D.7}$$

To calculate $\langle\langle \hat{c}_{k\sigma}, \hat{f}_\sigma^\dagger \rangle\rangle_\omega$ we again use the equation of motion Eq. (D.1), but in this case $\hat{A} = \hat{c}_{k\sigma}$ and $\hat{B} = \hat{f}_\sigma^\dagger$. Putting Eq. (D.5) into the equation of motion Eq. (D.1) gives the following relation:

$$(i\omega_n - \varepsilon_{k\sigma})\langle\langle \hat{c}_{k\sigma}, \hat{f}_\sigma^\dagger \rangle\rangle_\omega - V_{k\sigma}\langle\langle \hat{f}_\sigma, \hat{f}_\sigma^\dagger \rangle\rangle_\omega - \sigma W_k \langle\langle \hat{c}_{k\bar{\sigma}}^\dagger, \hat{f}_\sigma^\dagger \rangle\rangle_\omega = 0. \tag{D.8}$$

To calculate $\langle\langle \hat{c}_{k\bar{\sigma}}^\dagger, \hat{f}_\sigma^\dagger \rangle\rangle_\omega$ we once more use the equation of motion Eq. (D.1), but in this case $\hat{A} = \hat{c}_{k\bar{\sigma}}^\dagger$ and $\hat{B} = \hat{f}_\sigma^\dagger$. Putting Eq. (D.6) into the equation of motion Eq. (D.1) provides obtain the following relation:

$$(i\omega_n + \varepsilon_{k\bar{\sigma}})\langle\langle \hat{c}_{k\bar{\sigma}}^\dagger, \hat{f}_\sigma^\dagger \rangle\rangle_\omega + V_{k\bar{\sigma}}\langle\langle \hat{f}_{\bar{\sigma}}^\dagger, \hat{f}_\sigma^\dagger \rangle\rangle_\omega - \sigma W_k \langle\langle \hat{c}_{k\sigma}, \hat{f}_\sigma^\dagger \rangle\rangle_\omega = 0. \tag{D.9}$$

From Eqs. (D.8) and (D.9) we derive

$$\begin{aligned}\langle\langle \hat{c}_{k\sigma}, \hat{f}_\sigma^\dagger \rangle\rangle_\omega &= \frac{V_{k\sigma}(i\omega_n + \varepsilon_{k\bar{\sigma}})}{(i\omega_n - \varepsilon_{k\sigma})(i\omega_n + \varepsilon_{k\bar{\sigma}}) - W_k^2}\langle\langle \hat{f}_\sigma, \hat{f}_\sigma^\dagger \rangle\rangle_\omega \\ &- \frac{\sigma V_{k\bar{\sigma}} W_k}{(i\omega_n - \varepsilon_{k\sigma})(i\omega_n + \varepsilon_{k\bar{\sigma}}) - W_k^2}\langle\langle \hat{f}_{\bar{\sigma}}^\dagger, \hat{f}_\sigma^\dagger \rangle\rangle_\omega.\end{aligned} \tag{D.10}$$

Now we put Eq. (D.10) in Eq. (D.7) and we obtain:

$$\left(i\omega_n + \mu_{f\sigma} + \sum_k V_{k\sigma}^2 \frac{i\omega_n + \varepsilon_{k\bar{\sigma}}}{(\varepsilon_{k\sigma} - i\omega_n)(\varepsilon_{k\bar{\sigma}} + i\omega_n) + W_k^2}\right) \langle\langle \hat{f}_\sigma, \hat{f}_\sigma^\dagger \rangle\rangle_\omega$$
$$- \left(\sum_k \frac{\sigma V_{k\uparrow} V_{k\downarrow} W_k}{(\varepsilon_{k\sigma} - i\omega_n)(\varepsilon_{k\bar{\sigma}} + i\omega_n) + W_k^2}\right) \langle\langle \hat{f}_{\bar{\sigma}}^\dagger, \hat{f}_\sigma^\dagger \rangle\rangle_\omega$$
$$- U_f \langle\langle \hat{f}_\sigma \hat{f}_{\bar{\sigma}}^\dagger \hat{f}_{\bar{\sigma}}, \hat{f}_\sigma^\dagger \rangle\rangle_\omega - U_{fb} \langle\langle \hat{f}_\sigma \hat{b}^\dagger \hat{b}, \hat{f}_\sigma^\dagger \rangle\rangle_\omega + \sigma g \langle\langle \hat{f}_{\bar{\sigma}}^\dagger \hat{b}, \hat{f}_\sigma^\dagger \rangle\rangle_\omega = 1 \,. \quad \text{(D.11)}$$

Here we would like to mention that $\langle\langle \hat{f}_\sigma, \hat{f}_\sigma^\dagger \rangle\rangle_\omega \equiv G_\sigma(i\omega_n)$ is the normal Green's function and $\langle\langle \hat{f}_\uparrow, \hat{f}_\downarrow \rangle\rangle_\omega \equiv F(\omega)$ is the superconducting Green's function. We also define:

$$\Delta_\sigma(i\omega_n) = \Delta_\sigma^\star(-i\omega_n) = -\sum_k V_{k\sigma}^2 \frac{i\omega_n + \varepsilon_{k\bar{\sigma}}}{(\varepsilon_{k\sigma} - i\omega_n)(\varepsilon_{k\bar{\sigma}} + i\omega_n) + W_k^2} \,, \quad \text{(D.12)}$$

$$\Delta_{SC}(i\omega_n) = \Delta_{SC}^\star(-i\omega_n) = \sum_k \frac{V_{k\uparrow} V_{k\downarrow} W_k}{(\varepsilon_{k\uparrow} - i\omega_n)(\varepsilon_{k\downarrow} + i\omega_n) + W_k^2} \,, \quad \text{(D.13)}$$

the normal and the superconducting hybridization functions respectively and following correlation functions:

$$Q_{ff\sigma}(i\omega_n) = \langle\langle \hat{f}_\sigma \hat{f}_{\bar{\sigma}}^\dagger \hat{f}_{\bar{\sigma}}, \hat{f}_\sigma^\dagger \rangle\rangle_\omega \,, \qquad Q_{ff\sigma\bar{\sigma}}(i\omega_n) = \langle\langle \hat{f}_\sigma \hat{f}_{\bar{\sigma}}^\dagger \hat{f}_{\bar{\sigma}}, \hat{f}_{\bar{\sigma}} \rangle\rangle_\omega \,,$$
$$Q_{fb\sigma}(i\omega_n) = \langle\langle \hat{f}_\sigma \hat{b}^\dagger \hat{b}, \hat{f}_\sigma^\dagger \rangle\rangle_\omega \,, \qquad Q_{fb\sigma\bar{\sigma}}(i\omega_n) = \langle\langle \hat{f}_\sigma \hat{b}^\dagger \hat{b}, \hat{f}_{\bar{\sigma}} \rangle\rangle_\omega \,, \quad \text{(D.14)}$$
$$Q_{g\sigma}(i\omega_n) = \langle\langle \hat{f}_\sigma \hat{b}^\dagger, \hat{f}_\sigma^\dagger \rangle\rangle_\omega \,, \qquad Q_{g\sigma\bar{\sigma}}(i\omega_n) = \langle\langle \hat{f}_\sigma \hat{b}^\dagger, \hat{f}_{\bar{\sigma}} \rangle\rangle_\omega \,.$$

In this new definitions and using condition Eq. (C.11), Eq. (D.11) can be rewritten:

$$\left(i\omega_n + \mu_{f\sigma} - \Delta_\sigma(i\omega_n)\right) G_\sigma(i\omega_n) - \Delta_{SC}(\sigma i\omega_n) F^\star(-\sigma i\omega_n)$$
$$- U_f Q_{ff\sigma}(i\omega_n) - U_{fb} Q_{fb\sigma}(i\omega_n) - \sigma g Q_{g\bar{\sigma}\sigma}^\star(i\omega_n) = 1 \,. \quad \text{(D.15)}$$

To solve the problem we need to derive one more equation. For this purpose, we again use the equation of motion Eq. (D.1) and we consider that $\hat{A} = \hat{f}_\sigma^\dagger$ and $\hat{B} = \hat{f}_{\bar{\sigma}}^\dagger$. Based on Eq. (D.4) we get:

$$(i\omega_n - \mu_{f\sigma}) \langle\langle \hat{f}_\sigma^\dagger, \hat{f}_{\bar{\sigma}}^\dagger \rangle\rangle_\omega + U_f \langle\langle \hat{f}_\sigma^\dagger \hat{f}_{\bar{\sigma}}^\dagger \hat{f}_{\bar{\sigma}}, \hat{f}_{\bar{\sigma}}^\dagger \rangle\rangle_\omega + U_{fb} \langle\langle \hat{f}_\sigma^\dagger \hat{b}^\dagger \hat{b}, \hat{f}_{\bar{\sigma}}^\dagger \rangle\rangle_\omega$$
$$- \sigma g \langle\langle \hat{f}_{\bar{\sigma}} \hat{b}^\dagger, \hat{f}_{\bar{\sigma}}^\dagger \rangle\rangle_\omega + \sum_k V_{k\sigma} \langle\langle \hat{c}_{k\sigma}^\dagger, \hat{f}_{\bar{\sigma}}^\dagger \rangle\rangle_\omega = 0 \,. \quad \text{(D.16)}$$

We now have to calculate $\langle\langle \hat{c}_{k\sigma}^\dagger, \hat{f}_{\bar{\sigma}}^\dagger \rangle\rangle_\omega$. For this purpose we use Eqs. (D.8) and (D.9). So we obtain:

$$\langle\langle \hat{c}_{k\sigma}^{\dagger}, \hat{f}_{\bar{\sigma}}^{\dagger}\rangle\rangle_{\omega} = -\frac{\sigma V_{k\bar{\sigma}} W_k}{(i\omega_n - \varepsilon_{k\bar{\sigma}})(i\omega_n + \varepsilon_{k\sigma}) - W_k^2}\langle\langle \hat{f}_{\bar{\sigma}}, \hat{f}_{\bar{\sigma}}^{\dagger}\rangle\rangle_{\omega}$$
$$- \frac{V_{k\sigma}(i\omega_n - \varepsilon_{k\bar{\sigma}})}{(i\omega_n - \varepsilon_{k\bar{\sigma}})(i\omega_n + \varepsilon_{k\sigma}) - W_k^2}\langle\langle \hat{f}_{\sigma}^{\dagger}, \hat{f}_{\bar{\sigma}}^{\dagger}\rangle\rangle_{\omega}. \quad (D.17)$$

Putting Eq. (D.17) in the Eq. (D.16) we obtain:

$$\left(i\omega_n - \mu_{f\sigma} + \sum_k V_{k\sigma}^2 \frac{i\omega_n - \varepsilon_{k\bar{\sigma}}}{(\varepsilon_{k\bar{\sigma}} - i\omega_n)(\varepsilon_{k\sigma} + i\omega_n) + W_k^2}\right)\langle\langle \hat{f}_{\sigma}^{\dagger}, \hat{f}_{\bar{\sigma}}^{\dagger}\rangle\rangle_{\omega}$$
$$+ \left(\sum_k \frac{\sigma V_{k\uparrow} V_{k\downarrow} W_k}{(\varepsilon_{k\bar{\sigma}} - i\omega_n)(\varepsilon_{k\sigma} + i\omega_n) + W_k^2}\right)\langle\langle \hat{f}_{\bar{\sigma}}, \hat{f}_{\bar{\sigma}}^{\dagger}\rangle\rangle_{\omega}$$
$$+ U_f\langle\langle \hat{f}_{\sigma}^{\dagger} \hat{f}_{\bar{\sigma}}^{\dagger} \hat{f}_{\bar{\sigma}}, \hat{f}_{\bar{\sigma}}^{\dagger}\rangle\rangle_{\omega} + U_{fb}\langle\langle \hat{f}_{\sigma}^{\dagger} \hat{b}^{\dagger} \hat{b}, \hat{f}_{\bar{\sigma}}^{\dagger}\rangle\rangle_{\omega} - \sigma g\langle\langle \hat{f}_{\bar{\sigma}} \hat{b}^{\dagger}, \hat{f}_{\bar{\sigma}}^{\dagger}\rangle\rangle_{\omega} = 0. \quad (D.18)$$

Using definitions Eqs. (D.12), (D.13) and (D.14), one can rewrite Eq. (D.18) as follows:

$$-\sigma\left(i\omega_n - \mu_{f\sigma} + \Delta_{\sigma}^{\star}(i\omega_n)\right) F^{\star}(\sigma i\omega_n) + \sigma \Delta_{SC}(-\sigma i\omega_n) G_{\bar{\sigma}}(i\omega_n)$$
$$- U_f Q_{ff,\sigma\bar{\sigma}}^{\star}(i\omega_n) - U_{fb} Q_{fb\sigma\bar{\sigma}}^{\star}(i\omega_n) - \sigma g Q_{g\bar{\sigma}}(i\omega_n) = 0. \quad (D.19)$$

Now we write our results in the matrix form. For this purpose for $\sigma = 1$ we take Eq. (D.15) and complex conjugate of Eq. (D.19), while for $\sigma = -1$ we take Eq. (D.19) and complex conjugate of Eq. (D.15):

$$\left(i\omega_n + \mu_{f\uparrow} - \Delta_{\uparrow}(i\omega_n)\right) G_{\uparrow}(i\omega_n) - \Delta_{SC}(i\omega_n) F^{\star}(-i\omega_n)$$
$$- U_f Q_{ff\uparrow}(i\omega_n) - U_{fb} Q_{fb\uparrow}(i\omega_n) - g Q_{g\downarrow\uparrow}^{\star}(i\omega_n) = 1, \quad (D.20)$$
$$-\left(i\omega_n - \mu_{f\downarrow} + \Delta_{\downarrow}(-i\omega_n)\right) G_{\downarrow}^{\star}(i\omega_n) - \Delta_{SC}(i\omega_n) F(i\omega_n)$$
$$- U_f Q_{ff\downarrow}^{\star}(i\omega_n) - U_{fb} Q_{fb\downarrow}^{\star}(i\omega_n) + g Q_{g\uparrow\downarrow}(i\omega_n) = 1, \quad (D.21)$$
$$\left(i\omega_n + \mu_{f\uparrow} - \Delta_{\uparrow}(i\omega_n)\right) F(i\omega_n) + \Delta_{SC}(i\omega_n) G_{\downarrow}^{\star}(i\omega_n)$$
$$- U_f Q_{ff,\uparrow\downarrow}(i\omega_n) - U_{fb} Q_{fb\uparrow\downarrow}(i\omega_n) - g Q_{g\downarrow}^{\star}(i\omega_n) = 0, \quad (D.22)$$
$$\left(i\omega_n - \mu_{f\downarrow} + \Delta_{\downarrow}(i\omega_n)\right) F^{\star}(i\omega_n) - \Delta_{SC}(i\omega_n) G_{\uparrow}(i\omega_n)$$
$$- U_f Q_{ff,\downarrow\uparrow}^{\star}(i\omega_n) - U_{fb} Q_{fb\downarrow\uparrow}^{\star}(i\omega_n) + g Q_{g\uparrow}(i\omega_n) = 0. \quad (D.23)$$

The last four equations can be rewritten in the matrix form in a following way:

$$\begin{pmatrix} 1 & 0 \\ 0 & 1 \end{pmatrix} = \begin{pmatrix} i\omega_n + \mu_{f\uparrow} - \Delta_\uparrow(i\omega_n) & -\Delta_{SC}(i\omega_n) \\ -\Delta_{SC}(i\omega_n) & i\omega_n - \mu_{f\downarrow} + \Delta_\downarrow(-i\omega_n) \end{pmatrix} \begin{pmatrix} G_\uparrow(i\omega_n) & F(i\omega_n) \\ F^\star(-i\omega_n) & -G^\star_\downarrow(i\omega_n) \end{pmatrix} \quad \text{(D.24)}$$
$$- \begin{pmatrix} U_f Q_{ff\uparrow}(i\omega_n) + U_{fb} Q_{fb\uparrow}(i\omega_n) + gQ^\star_{g\downarrow\uparrow}(i\omega_n) & U_f Q_{ff,\uparrow\downarrow}(i\omega_n) + U_{fb} Q_{fb\uparrow\downarrow}(i\omega_n) + gQ^\star_{g\downarrow}(i\omega_n) \\ U_f Q^\star_{ff,\downarrow\uparrow}(i\omega_n) + U_{fb} Q^\star_{fb\downarrow\uparrow}(i\omega_n) - gQ_{g\uparrow}(i\omega_n) & U_f Q^\star_{ff\downarrow}(i\omega_n) + U_{fb} Q^\star_{fb\downarrow}(i\omega_n) - gQ_{g\uparrow\downarrow}(i\omega_n) \end{pmatrix}.$$

Now we compare Eq. (D.24) with the Dyson equation, which now have the matrix form:

$$\hat{\mathcal{G}}^{-1}(i\omega_n) - \hat{\Sigma}(i\omega_n) = \hat{G}^{-1}(i\omega_n), \quad \text{(D.25)}$$

where

$$\hat{G}(i\omega) = \begin{pmatrix} G_\uparrow(i\omega_n) & F(i\omega_n) \\ F^\star(-i\omega_n) & -G^\star_\downarrow(i\omega_n) \end{pmatrix} \quad \text{(D.26)}$$

is the matrix interacting Green's function,

$$\hat{\mathcal{G}}(i\omega) = \begin{pmatrix} i\omega_n + \mu_{f\uparrow} - \Delta_\uparrow(i\omega_n) & -\Delta_{SC}(i\omega_n) \\ -\Delta_{SC}(i\omega_n) & i\omega_n - \mu_{f\downarrow} + \Delta_\downarrow(-i\omega_n) \end{pmatrix}^{-1} \quad \text{(D.27)}$$

is the matrix Weiss Green's function and $\hat{\Sigma}(\omega)$ is the matrix self-energy. From this comparison directly follows that

$$\begin{pmatrix} \Sigma_\uparrow(i\omega_n) & \Sigma_{SC}(i\omega_n) \\ \Sigma^\star_{SC}(-i\omega_n) & -\Sigma^\star_\downarrow(i\omega_n) \end{pmatrix}$$
$$= \begin{pmatrix} U_f Q_{ff\uparrow}(i\omega_n) + U_{fb} Q_{fb\uparrow}(i\omega_n) + gQ^\star_{g\downarrow\uparrow}(i\omega_n) & U_f Q_{ff,\uparrow\downarrow}(i\omega_n) + U_{fb} Q_{fb\uparrow\downarrow}(i\omega_n) + gQ^\star_{g\downarrow}(i\omega_n) \\ U_f Q^\star_{ff,\downarrow\uparrow}(i\omega_n) + U_{fb} Q^\star_{fb\downarrow\uparrow}(i\omega_n) - gQ_{g\uparrow}(i\omega_n) & U_f Q^\star_{ff\downarrow}(i\omega_n) + U_{fb} Q^\star_{fb\downarrow}(i\omega_n) - gQ_{g\uparrow\downarrow}(i\omega_n) \end{pmatrix}$$
$$\times \begin{pmatrix} G_\uparrow(i\omega_n) & F(i\omega_n) \\ F^\star(-i\omega_n) & -G^\star_\downarrow(i\omega_n) \end{pmatrix}^{-1}. \quad \text{(D.28)}$$

From here we directly obtain that:

$$\Sigma_\sigma(i\omega_n) = \frac{\left(U_f Q_{ff\sigma}(i\omega_n) + U_{fb} Q_{fb\sigma}(i\omega_n) + \sigma g Q^\star_{g\bar{\sigma}\sigma}(i\omega_n)\right) G^\star_{\bar{\sigma}}(i\omega_n)}{G_\sigma(i\omega_n) G^\star_{\bar{\sigma}}(i\omega_n) + F(\sigma i\omega_n) F^\star(\bar{\sigma} i\omega_n)}$$
$$+ \frac{\left(\sigma U_f Q_{ff,\sigma\bar{\sigma}}(i\omega_n) + \sigma U_{fb} Q_{fb\sigma\bar{\sigma}}(i\omega_n) + g Q^\star_{g\bar{\sigma}}(i\omega_n)\right) F^\star(\bar{\sigma} i\omega_n)}{G_\sigma(i\omega_n) G^\star_{\bar{\sigma}}(i\omega_n) + F(\sigma i\omega_n) F^\star(\bar{\sigma} i\omega_n)}, \quad \text{(D.29)}$$

$$\Sigma_{SC}(i\omega_n) = \frac{\left(U_f Q_{ff\uparrow}(i\omega_n) + U_{fb} Q_{fb\uparrow}(i\omega_n) + g Q^\star_{g\downarrow\uparrow}(i\omega_n)\right) F(i\omega_n)}{G_\uparrow(i\omega_n) G^\star_\downarrow(i\omega_n) + F(i\omega_n) F^\star(-i\omega_n)}$$
$$- \frac{\left(U_f Q_{ff,\uparrow\downarrow}(i\omega_n) + U_{fb} Q_{fb\uparrow\downarrow}(i\omega_n) + g Q^\star_{g\downarrow}(i\omega_n)\right) G_\uparrow(i\omega_n)}{G_\uparrow(i\omega_n) G^\star_\downarrow(i\omega_n) + F(i\omega_n) F^\star(-i\omega_n)}, \quad \text{(D.30)}$$

$$\Sigma^\star_{SC}(-i\omega_n) = \frac{\left(U_f Q^\star_{ff\downarrow}(i\omega_n) + U_{fb} Q^\star_{fb\downarrow}(i\omega_n) - g Q_{g\uparrow\downarrow}(i\omega_n)\right) F^\star(-i\omega_n)}{G_\uparrow(i\omega_n) G^\star_\downarrow(i\omega_n) + F(i\omega_n) F^\star(-i\omega_n)}$$
$$+ \frac{\left(U_f Q^\star_{ff,\downarrow\uparrow}(i\omega_n) + U_{fb} Q^\star_{fb\downarrow\uparrow}(i\omega_n) - g Q_{g\uparrow}(i\omega_n)\right) G^\star_\downarrow(i\omega_n)}{G_\uparrow(i\omega_n) G^\star_\downarrow(i\omega_n) + F(i\omega_n) F^\star(-i\omega_n)}. \quad \text{(D.31)}$$

Appendix E

Derivation of the Kinetic Energy

In this appendix we derive the equation for the fermionic kinetic energy for system with a two sub-lattice structure. It is well known that the fermionic kinetic energy is given by (to simplify the notations, we drop the summation over the spin index σ):

$$\hat{\mathcal{E}}_{kin} = -t \sum_{\langle ij \rangle} \hat{c}_i^\dagger \hat{c}_j \,, \tag{E.1}$$

where $\langle ij \rangle$ means summation over nearest neighbors. We now introduce the fermionic creation operators in the energy eigenbasis:

$$\hat{c}_n = \frac{1}{\sqrt{N}} \sum_i X_{ni} \hat{c}_i \,, \tag{E.2}$$

where N is the number of lattice sites. The inverse transformation has the following form:

$$\hat{c}_i = \frac{1}{\sqrt{N}} \sum_n X_{in}^\star \hat{c}_n \,. \tag{E.3}$$

The following condition ensures that after the transformation the Hamiltonian becomes diagonal:

$$-\frac{t}{N} \sum_{\langle ij \rangle} X_{ni} X_{jn'}^\star = -\frac{t}{N} \sum_{+\langle ij \rangle_-} \left(X_{ni} X_{jn'}^\star + X_{nj} X_{in'}^\star \right) = -\frac{2t}{N} \sum_{+\langle ij \rangle_-} X_{ni} X_{jn'}^\star = \delta_{nn'} \varepsilon_n \,, \tag{E.4}$$

where $_\alpha \langle ij \rangle_{\bar{\alpha}}$ denotes summation over the nearest neighbors such that i belongs to the sublattice α and j belongs to the sublattice $\bar{\alpha} = -\alpha$. At this point we have assumed that the lattice is bipartite. The second equality is based on the fact that both sublattices are identical and therefore $\sum_{+\langle ij\rangle_-} = \sum_{-\langle ij\rangle_+}$.

For a bipartite lattice one can reverse the sign of the fermion creation/annihilation operators on one of the sublattices. This again yields an eigenstate of the Hamiltonian (E.1), but with opposite sign. From this it directly follows that for each single-particle state with energy ε_n, there exists a state with energy $-\varepsilon_n$, i.e we can label the eigenstates such that

$$\varepsilon_{n+N/2} = -\varepsilon_n. \tag{E.5}$$

From the Eqs. (E.4) and (E.5) it then follows that:

$$X_{i \in S_1, n+N/2} = X_{in} \quad \text{and} \quad X_{j \in S_{-1}, n+N/2} = -X_{jn}, \tag{E.6}$$

where S_α ($\alpha = \pm 1$) denotes the set of lattice points in sublattice α.

Now we introduce two new operators

$$\hat{c}_{n,1} = \frac{1}{\sqrt{2}} \left(\hat{c}_n + \hat{c}_{n+N/2} \right) = \frac{1}{\sqrt{N/2}} \sum_{i \in S_1} X_{ni} \hat{c}_i, \tag{E.7}$$

$$\hat{c}_{n,-1} = \frac{1}{\sqrt{2}} \left(\hat{c}_n - \hat{c}_{n+N/2} \right) = \frac{1}{\sqrt{N/2}} \sum_{j \in S_{-1}} X_{nj} \hat{c}_j. \tag{E.8}$$

Here and later we work modulo N, i.e. $n + N = n$. From Eqs. (E.7) and (E.8) one easily obtains the following identity:

$$\hat{c}_{n+N/2,\pm 1} = \pm \hat{c}_{n,\pm 1}. \tag{E.9}$$

The inverse transformation has the following form:

$$\hat{c}_{i \in S_1} = \frac{1}{\sqrt{N/2}} \sum_{n=1}^{N/2} X_{in}^\star \hat{c}_{n,1}, \tag{E.10}$$

$$\hat{c}_{j \in S_{-1}} = \frac{1}{\sqrt{N/2}} \sum_{n=1}^{N/2} X_{jn}^\star \hat{c}_{n,-1}. \tag{E.11}$$

Using Eqs. (E.1), (E.4), (E.5), (E.9), (E.10) and (E.11) we obtain

$$
\begin{aligned}
\hat{\mathcal{E}}_{kin} &= -t \sum_{+\langle ij \rangle_-} \left(\hat{c}_i^\dagger \hat{c}_j + \hat{c}_j^\dagger \hat{c}_i \right) = -t \sum_{+\langle ij \rangle_-} \sum_{n,n'}^{N/2} \left(\frac{1}{N/2} X_{ni} X_{jn'}^\star \hat{c}_{n,1}^\dagger \hat{c}_{n',-1} + h.c \right) \\
&= \sum_{n,n'}^{N/2} \left[\left(-\frac{2t}{N} \sum_{+\langle ij \rangle_-} X_{ni} X_{jn'}^\star \right) \hat{c}_{n,1}^\dagger \hat{c}_{n',-1} + h.c \right] = \sum_{n=1}^{N/2} \varepsilon_n \left(\hat{c}_{n,1}^\dagger \hat{c}_{n,-1} + h.c \right) \\
&= \frac{1}{2} \sum_{n=1}^{N} \varepsilon_n \left(\hat{c}_{n,1}^\dagger \hat{c}_{n,-1} + h.c \right) .
\end{aligned} \quad (E.12)
$$

In the last step we have used (E.5) and (E.9) as follows:

$$
\sum_{n=1}^{N/2} \varepsilon_n \hat{c}_{n,1}^\dagger \hat{c}_{n,-1} = \sum_{n=1}^{N/2} (-\varepsilon_{n+N/2}) \hat{c}_{n+N/2,1}^\dagger (-\hat{c}_{n+N/2,-1}) = \sum_{n=N/2+1}^{N} \varepsilon_n \hat{c}_{n,1}^\dagger \hat{c}_{n,-1} . \quad (E.13)
$$

The next step is to go from summation to integral, and to take the expectation value of the kinetic energy operator. We obtain:

$$
\begin{aligned}
\mathcal{E}_{kin} &= \frac{1}{2} \left\langle \int_{-\infty}^{\infty} d\varepsilon\, \rho(\varepsilon)\varepsilon \left(\hat{c}_{\varepsilon,1}^\dagger \hat{c}_{\varepsilon,-1} + h.c \right) \right\rangle \\
&= \lim_{\tau \to 0} \frac{1}{2} \int_{-\infty}^{\infty} d\varepsilon\, \rho(\varepsilon)\varepsilon \left(\langle \hat{c}_{\varepsilon,1}^\dagger(0) \hat{c}_{\varepsilon,-1}(\tau) \rangle + \langle \hat{c}_{\varepsilon,-1}^\dagger(0) \hat{c}_{\varepsilon,1}(\tau) \rangle \right) \\
&= \lim_{\tau \to 0} \int_{-\infty}^{\infty} d\varepsilon\, \rho(\varepsilon)\varepsilon \mathcal{B}(\varepsilon,\tau) = \lim_{\tau \to 0} k_B T \int_{-\infty}^{\infty} d\varepsilon\, \rho(\varepsilon)\varepsilon \sum_n e^{-i\omega_n \tau} \mathcal{B}(\varepsilon,\omega_n) \\
&= k_B T \sum_n \int_{-\infty}^{\infty} d\varepsilon\, \rho(\varepsilon)\varepsilon \mathcal{B}(\varepsilon,\omega_n) = \int_{-\infty}^{\infty} d\varepsilon\, \rho(\varepsilon)\varepsilon \int_{-\infty}^{\infty} d\omega f(\omega) B(\varepsilon,\omega^+) , \quad (E.14)
\end{aligned}
$$

where $\mathcal{B}(\varepsilon,\tau) = \frac{1}{2} \left(\langle \hat{c}_{\varepsilon,1}^\dagger(0) \hat{c}_{\varepsilon,-1}(\tau) \rangle + \langle \hat{c}_{\varepsilon,-1}^\dagger(0) \hat{c}_{\varepsilon,1}(\tau) \rangle \right)$ and $B = -\frac{1}{\pi} \text{Im} \mathcal{B}$

These two terms are just the off-diagonal terms of the following Green's function matrix, which according to the Dyson equation has the form:

$$
\begin{aligned}
\hat{G}^{-1}(\varepsilon,\omega_n) &= \begin{pmatrix} i\omega_n + \mu_f & -\varepsilon \\ -\varepsilon & i\omega_n + \mu_f \end{pmatrix} - \begin{pmatrix} \Sigma_1(\omega) & 0 \\ 0 & \Sigma_{-1}(\omega) \end{pmatrix} \\
&= \begin{pmatrix} i\omega_n + \mu_f - \Sigma_1 & -\varepsilon \\ -\varepsilon & i\omega_n + \mu_f - \Sigma_{-1} \end{pmatrix} .
\end{aligned} \quad (E.15)
$$

We obtain

$$\hat{G}(\varepsilon,\omega_n) = \begin{pmatrix} \frac{\zeta_{-1}}{\zeta_1\zeta_{-1}-\varepsilon^2} & \frac{\varepsilon}{\zeta_1\zeta_{-1}-\varepsilon^2} \\ \frac{\varepsilon}{\zeta_1\zeta_{-1}-\varepsilon^2} & \frac{\zeta_{-1}}{\zeta_1\zeta_{-1}-\varepsilon^2} \end{pmatrix}, \qquad (E.16)$$

where

$$\zeta_\alpha(\omega_n) = i\omega_n + \mu - \Sigma_\alpha. \qquad (E.17)$$

Therefore

$$\mathcal{B}(\varepsilon,\omega_n) = \frac{\varepsilon}{\zeta_1\zeta_{-1}-\varepsilon^2} = \frac{1}{2}\left(\frac{1}{\sqrt{\zeta_1\zeta_{-1}}-\varepsilon} - \frac{1}{\sqrt{\zeta_1\zeta_{-1}}+\varepsilon}\right). \qquad (E.18)$$

As one can easily see, the integral in Eq. (E.14) remains invariant if we replace $\mathcal{B}(\varepsilon,\omega_n)$ by the following expression:

$$\mathcal{B}(\varepsilon,\omega_n) = \frac{1}{\sqrt{\zeta_1\zeta_{-1}}-\varepsilon}. \qquad (E.19)$$

The advantage of this representation is that in the limit of one-sublattice it reduces to the "usual" equation of the spectral function.

Appendix F

Iterative Diagonalization within NRG

In this appendix we review in detail iterative diagonalization within NRG. As we have mentioned in subsection 3.4.2 the iterative diagonalization method based on the fact that:

$$\hat{\mathcal{H}}_{N+1} = \Lambda^{1/2}\hat{\mathcal{H}}_N + \Lambda^{N/2}\sum_\sigma t_{N\sigma}(\hat{d}^\dagger_{N\sigma}\hat{d}_{N+1,\sigma} + h.c) + \Lambda^{N/2}\sum_\sigma \delta_{N+1,\sigma}\hat{d}^\dagger_{N+1,\sigma}\hat{d}_{N+1,\sigma}. \quad (F.1)$$

Let's imagine that we know all eigenvectors and eigenvalues and all matrix elements of the Hamiltonian $\hat{\mathcal{H}}_N$. Then we can construct new basis states for the Hamiltonian $\hat{\mathcal{H}}_{N+1}$:

$$\begin{aligned}
|1, r, Q, S_z\rangle_{N+1} &= |r, Q+1, S_z\rangle_N, \\
|2, r, Q, S_z\rangle_{N+1} &= \hat{d}^\dagger_{N+1,\uparrow}|r, Q, S_z - 1/2\rangle_N, \\
|3, r, Q, S_z\rangle_{N+1} &= \hat{d}^\dagger_{N+1,\downarrow}|r, Q, S_z + 1/2\rangle_N, \\
|4, r, Q, S_z\rangle_{N+1} &= \hat{d}^\dagger_{N+1,\uparrow}\hat{d}^\dagger_{N+1,\downarrow}|r, Q-1, S_z\rangle_N,
\end{aligned} \quad (F.2)$$

where

$$Q = \sum_{\sigma,n=0}^N (\hat{d}^\dagger_{n,\sigma}\hat{d}_{n,\sigma} - 1) + \sum_\sigma (\hat{f}^\dagger_\sigma \hat{f}_\sigma - 1) \quad \text{and} \quad S_z = \frac{1}{2}\sum_{\sigma,n=0}^N (\hat{d}^\dagger_{n,\uparrow}\hat{d}_{n,\uparrow} - \hat{d}^\dagger_{n,\downarrow}\hat{d}_{n,\downarrow}) + \frac{1}{2}(\hat{f}^\dagger_\uparrow \hat{f}_\uparrow - \hat{f}^\dagger_\downarrow \hat{f}_\downarrow)$$

are conserving quantum numbers describing number and the spin of the fermions respectively. r labels different energy levels with quantum numbers Q and S_z.

Now as we know the new basis states for the Hamiltonian $\hat{\mathcal{H}}_{N+1}$, we can build up the matrix elements of the Hamiltonian $\hat{\mathcal{H}}_{N+1}$. The diagonal elements of the new Hamiltonian are:

$$\langle 1,r,Q,S_z|\hat{\mathcal{H}}_{N+1}|1,r,Q,S_z\rangle_{N+1} = \sqrt{\Lambda}\langle r,Q+1,S_z|\hat{\mathcal{H}}_N|r,Q+1,S_z\rangle_N = \sqrt{\Lambda}E_r(Q+1,S_z),$$
$$\langle 2,r,Q,S_z|\hat{\mathcal{H}}_{N+1}|2,r,Q,S_z\rangle_{N+1} = \sqrt{\Lambda}\langle r,Q,S_z-1/2|\hat{\mathcal{H}}_N|r,Q,S_z-1/2\rangle_N + \Lambda^{N/2}\delta_{N+1,\uparrow}$$
$$= \sqrt{\Lambda}E_r(Q,S_z-1/2) + \Lambda^{N/2}\delta_{N+1,\uparrow}, \quad (F.3)$$
$$\langle 3,r,Q,S_z|\hat{\mathcal{H}}_{N+1}|3,r,Q,S_z\rangle_{N+1} = \sqrt{\Lambda}\langle r,Q,S_z+1/2|\hat{\mathcal{H}}_N|r,Q,S_z+1/2\rangle_N + \Lambda^{N/2}\delta_{N+1,\downarrow}$$
$$= \sqrt{\Lambda}E_r(Q,S_z+1/2) + \Lambda^{N/2}\delta_{N+1,\downarrow},$$
$$\langle 4,r,Q,S_z|\hat{\mathcal{H}}_{N+1}|4,r,Q,S_z\rangle_{N+1} = \sqrt{\Lambda}\langle r,Q-1,S_z|\hat{\mathcal{H}}_N|r,Q-1,S_z\rangle_N + \delta_{N+1,\uparrow} + \delta_{N+1,\downarrow}$$
$$= \sqrt{\Lambda}E_r(Q-1,S_z) + \Lambda^{N/2}\delta_{N+1,\uparrow} + \Lambda^{N/2}\delta_{N+1,\downarrow}.$$

We calculate the off-diagonal terms of the Hamiltonian matrix:

$$\langle 2,r',Q,S_z|\hat{\mathcal{H}}_{N+1}|1,r,Q,S_z\rangle_{N+1} = \Lambda^N t_{N\uparrow}\langle r,Q+1,S_z|\hat{d}^\dagger_{N\uparrow}|r',Q,S_z-1/2\rangle_N,$$
$$\langle 4,r',Q,S_z|\hat{\mathcal{H}}_{N+1}|3,r,Q,S_z\rangle_{N+1} = -\Lambda^N t_{N\uparrow}\langle r,Q,S_z+1/2|\hat{d}^\dagger_{N\uparrow}|r',Q-1,S_z\rangle_N \quad (F.4)$$
$$\langle 3,r',Q,S_z|\hat{\mathcal{H}}_{N+1}|1,r,Q,S_z\rangle_{N+1} = \Lambda^N t_{N\downarrow}\langle r,Q+1,S_z|\hat{d}^\dagger_{N\downarrow}|r',Q,S_z+1/2\rangle_N,$$
$$\langle 4,r',Q,S_z|\hat{\mathcal{H}}_{N+1}|2,r,Q,S_z\rangle_{N+1} = \Lambda^N t_{N\downarrow}\langle r,Q,S_z-1/2|\hat{d}^\dagger_{N\downarrow}|r',Q-1,S_z\rangle_N.$$

Using Eqs. (F.3) and (F.4) we can fill the Hamiltonian matrix. Afterwards we can diagonalize it and find new eigenvalues and eigenvectors. As already discussed in the subsection (3.4.2), the Hamiltonian matrix is block diagonal, so we can independently diagonalize each block of the Hamiltonian matrix, which can be characterized by quantum numbers Q and S_z. We define the transformations matrices by $U_{l;i,r}(Q,S,z)$, then we can express the new states by basis states which we defined in (F.2) as follows:

$$|l,Q,S_z\rangle_{N+1} = \sum_{i,r} U_{l;i,r}(Q,S,z)|i,r,Q,S_z\rangle_N. \quad (F.5)$$

Now we can construct the $\hat{\mathcal{H}}_{N+2}$ Hamiltonian, for this purpose we have to calculate the following matrix elements:

$$\langle r', Q, S_z | \hat{d}^\dagger_{N+1,\uparrow} | r, Q, S_z \rangle_{N+1} = \sum_{l \in (Q+1, S_z)} U_{r';2,l}(Q+1, S, z+1/2) U_{r;1,l}(Q, S, z)$$
$$+ \sum_{l \in (Q, S_z+1/2)} U_{r';4,l}(Q+1, S, z+1/2) U_{r;3,l}(Q, S, z), \quad \text{(F.6)}$$

$$\langle r', Q, S_z | \hat{d}^\dagger_{N+1,\downarrow} | r, Q, S_z \rangle_{N+1} = \sum_{l \in (Q+1, S_z)} U_{r';3,l}(Q+1, S, z-1/2) U_{r;1,l}(Q, S, z)$$
$$- \sum_{l \in (Q, S_z-1/2)} U_{r';4,l}(Q+1, S, z-1/2) U_{r;2,l}(Q, S, z). \quad \text{(F.7)}$$

So this allows us to calculate new matrix elements for the Hamiltonian $\hat{\mathcal{H}}_{N+1}$, and now we can build the Hamiltonian $\hat{\mathcal{H}}_{N+2}$ and continue this processes until the desired number of lattice sites in the linear chain is reached.

In the end of this Appendix, we show how different correlation functions are transformed during the iterative diagonalization processes. In particular, we consider two different types of operators: (i) $\hat{\mathcal{O}}_{\pm 1}$, which increases/decreases the number of fermions in the impurity site, and (ii) $\hat{\mathcal{O}}_0$, which does not change the number of fermions in the impurity site.

$$\langle r', Q+1, S_z + \frac{1}{2}\sigma | \hat{\mathcal{O}}_\sigma | r, Q, S_z \rangle_{N+1} \quad \text{(F.8)}$$
$$= \sum_{l,l',i} U_{r';i,l'}(Q+1, S, z+\frac{1}{2}\sigma) U_{r;i,l}(Q, S, z) \langle i, l', Q+1, S_z + \frac{1}{2}\sigma | \hat{\mathcal{O}}_\sigma | i, l, Q, S_z \rangle_{N+1},$$
$$\langle r', Q, S_z | \hat{\mathcal{O}}_0 | r, Q, S_z \rangle_{N+1} \quad \text{(F.9)}$$
$$= \sum_{l,l',i} U_{r';i,l'}(Q, S, z) U_{r;i,l}(Q, S, z) \langle i, l', Q, S_z | \hat{\mathcal{O}}_0 | i, l, Q, S_z \rangle_{N+1}.$$

The last step is to express everything by correlation functions after N iteration:

$$\langle 1, l', Q+1, S_z + \tfrac{1}{2}\sigma | \hat{\mathcal{O}}_\sigma | 1, l, Q, S_z \rangle_{N+1} = \langle l', Q+2, S_z + \tfrac{1}{2}\sigma | \hat{\mathcal{O}}_\sigma | l, Q+1, S_z \rangle_N,$$

$$\langle 2, l', Q+1, S_z + \tfrac{1}{2}\sigma | \hat{\mathcal{O}}_\sigma | 2, l, Q, S_z \rangle_{N+1} = -\langle l', Q+1, S_z + \tfrac{1}{2}(\sigma-1) | \hat{\mathcal{O}}_\sigma | l, Q+1, S_z - \tfrac{1}{2} \rangle_N,$$

$$\langle 3, l', Q+1, S_z + \tfrac{1}{2}\sigma | \hat{\mathcal{O}}_\sigma | 3, l, Q, S_z \rangle_{N+1} = -\langle l', Q+1, S_z + \tfrac{1}{2}(\sigma+1) | \hat{\mathcal{O}}_\sigma | l, Q+1, S_z + \tfrac{1}{2} \rangle_N,$$

$$\langle 4, l', Q+1, S_z + \tfrac{1}{2}\sigma | \hat{\mathcal{O}}_\sigma | 4, l, Q, S_z \rangle_{N+1} = \langle l', Q, S_z + \tfrac{1}{2}\sigma | \hat{\mathcal{O}}_\sigma | l, Q-1, S_z \rangle_N,$$

$$\langle 1, l', Q, S_z | \hat{\mathcal{O}}_0 | 1, l, Q, S_z \rangle_{N+1} = \langle l', Q+1, S_z | \hat{\mathcal{O}}_0 | i, l, Q+1, S_z \rangle_N,$$

$$\langle 2, l', Q, S_z | \hat{\mathcal{O}}_0 | 2, l, Q, S_z \rangle_{N+1} = \langle l', Q, S_z - \tfrac{1}{2} | \hat{\mathcal{O}}_0 | i, l, Q, S_z - \tfrac{1}{2} \rangle_N,$$

$$\langle 3, l', Q, S_z | \hat{\mathcal{O}}_0 | 3, l, Q, S_z \rangle_{N+1} = \langle l', Q, S_z + \tfrac{1}{2} | \hat{\mathcal{O}}_0 | i, l, Q, S_z + \tfrac{1}{2} \rangle_N,$$

$$\langle 4, l', Q, S_z | \hat{\mathcal{O}}_0 | 4, l, Q, S_z \rangle_{N+1} = \langle l', Q-1, S_z | \hat{\mathcal{O}}_0 | i, l, Q-1, S_z \rangle_N,$$

Bibliography

[1] M. Greiner, O. Mandel, T. Esslinger, T. W. Hänsch, and I. Bloch *Nature*, vol. 415, p. 39, 2002.

[2] D. Jaksch, C. Bruder, J. I. Cirac, C. W. Gardiner, and P. Zoller *Phys. Rev. Lett.*, vol. 81, p. 3108, 1998.

[3] W. Hofstetter, J. I. Cirac, P. Zoller, E. Demler, and M. D. Lukin *Phys. Rev. Lett.*, vol. 89, p. 220407, 2002.

[4] A. G. Truscott, K. E. Strecker, W. I. Mc Alexander, G. B. Partidge, and R. G. Hulet *Science*, vol. 291, p. 2570, 2001.

[5] Z. Hadzibabic, C. A. Stan, K. Dieckmann, S. Gupta, M. W. Zwierlein, A. Görlitz, and W. Ketterle *Phys. Rev. Lett.*, vol. 88, p. 160401, 2002.

[6] Z. Hadzibabic, S. Gupta, C. A. Stan, C. H. Schunck, M. W. Zwierlein, K. Dieckmann, and W. Ketterle *Phys. Rev. Lett.*, vol. 91, p. 160401, 2003.

[7] G. Roati, F. Riboli, G. Modugno, and M. Inguscio *Phys. Rev. Lett.*, vol. 89, p. 150403, 2002.

[8] C. Silber, S. Günther, C. Marzok, B. Deh, P. W. Courteille, and C. Zimmermann *Phys. Rev. Lett.*, vol. 95, p. 170408, 2005.

[9] M. Zaccanti, C. D'Errico, F. Ferlaino, G. Roati, M. Inguscio, and G. Modugno *Phys. Rev. A*, vol. 74, p. 041605R, 2006.

[10] D. S. Lühmann, K. Bongs, K. Sengstock, and D. Pfannkuche *Phys. Rev. Lett.*, vol. 101, p. 050402, 2008.

[11] M. Cramer, S. Ospelkaus, C. Ospelkaus, K. Bongs, K. Sengstock, and J. Eisert *Phys. Rev. Lett.*, vol. 100, p. 160409, 2008.

[12] S. Ospelkaus, C. Ospelkaus, O. Wille, M. Succo, P. Ernst, K. Sengstock, and K. Bongs *Phys. Rev. Lett.*, vol. 96, p. 180403, 2006.

[13] C. Ospelkaus, S. Ospelkaus, K. Sengstock, and K. Bongs *Phys. Rev. Lett.*, vol. 96, p. 020401, 2006.

[14] C. Ospelkaus, S. Ospelkaus, L. Humbert, P. Ernst, K. Sengstock, and K. Bongs *Phys. Rev. Lett.*, vol. 97, p. 120402, 2006.

[15] K. Günter, T. Stöferle, H. Moritz, M. Köhl, and T. Esslinger *Phys. Rev. Lett.*, vol. 96, p. 180402, 2006.

[16] T. Best, S. Will, U. Schneider, L. Hackermueller, D. S. Luehmann, D. van Oosten, and I. Bloch *Phys. Rev. Lett.*, vol. 102, p. 030408, 2009.

[17] F. Schreck, L. Khaykovich, K. L. Corwin, F. G., T. Bourdel, J. Cubizollles, and C. Salomon *Phys. Rev. Lett.*, vol. 87, p. 080403, 2001.

[18] M. Modugno, F. Ferlaino, F. Riboli, G. Roati, G. Modugno, and M. Inguscio *Phys. Rev. A*, vol. 68, p. 043626, 2003.

[19] F. Ferlaino, E. de Mirandes, G. Roati, G. Modugno, and M. Inguscio *Phys. Rev. Lett.*, vol. 92, p. 140405, 2004.

[20] I. Titvinidze, M. Snoek, and W. Hofstetter *Phys. Rev. Lett.*, vol. 100, p. 100401, 2008.

[21] I. Titvinidze, M. Snoek, and W. Hofstetter *Phys. Rev. B*, vol. 79, p. 144506, 2009.

[22] A. Albus, F. Illuminati, and J. Eisert *Phys. Rev A*, vol. 68, p. 023606, 2003.

[23] L. Pollet, M. Troyer, K. Van Houcke, and S. M. A. Rombouts *Phys. Rev. Lett.*, vol. 96, p. 190402, 2006.

[24] L. Mathey, S. W. Tsai, and A. H. Castro Neto *Phys. Rev. Lett.*, vol. 97, p. 030601, 2006.

[25] A. Imambekov and E. Demler *Phys. Rev A*, vol. 73, p. 021602R, 2006.

[26] L. Mathey, D. W. Wang, W. Hofstetter, M. D. Lukin, and E. Demler *Phys. Rev. Lett.*, vol. 93, p. 120404, 2004.

[27] R. Roth and K. Burnett *Phys. Rev A*, vol. 69, p. 021601, 2004.

[28] M. A. Cazalilla and A. F. Ho *Phys. Rev. Lett.*, vol. 91, p. 150403, 2003.

[29] H. P. Büchler and G. Blatter *Phys. Rev. Lett.*, vol. 91, p. 130404, 2003.

[30] M. Lewenstein, L. Santos, M. A. Baranov, and H. Fehrmann *Phys. Rev. Lett.*, vol. 92, p. 050401, 2004.

[31] F. D. Klironomos and S. W. Tsai *Phys. Rev. Lett.*, vol. 99, p. 100401, 2007.

[32] L. Pollet, C. Kollath, U. Schollwöck, and M. Troyer *Physical Review* A, vol. 77, p. 023608, 2008.

[33] G. Refael and E. Demler *Physical Review* B, vol. 77, p. 144511, 2008.

[34] R. M. Lutchyn, S. Tewari, and S. D. Sarma *Physical Review* B, vol. 78, p. 220504(R), 2008.

[35] A. Mering and M. Fleischhauer *Physical Review* A, vol. 77, p. 023601, 2008.

[36] X. Barillier-Pertuisel, S. Pittel, L. Pollet, and P. Schuck *Physical Review* A, vol. 77, p. 012115, 2008.

[37] F. Hébert, F. Haudin, L. Pollet, and G. G. Batrouni *Physical Review* A, vol. 76, p. 043619, 2007.

[38] A. Zujev, A. Baldwin, R. T. Scalettar, V. G. Rousseau, P. J. H. Denteneer, and M. Rigol *Physical Review* A, vol. 78, p. 033619, 2008.

[39] C. N. Varney, V. G. Rousseau, and R. T. Scalettar *Physical Review* A, vol. 77, p. 041608, 2008.

[40] A. W. Sandvik *Physical Review Letters*, vol. 101, p. 120405, 2008.

[41] W.-Q. Ning, S.-J. Gu, Y.-G. Chen, C.-Q. Wu, and H.-Q. Lin *Journal of Physics: Condensed Matter*, vol. 20, p. 235236, 2008.

[42] S. Roöthel and A. Pelster *European Physical Journal* B, vol. 59, p. 343, 2007.

[43] D. S. Luhmann, K. Bongs, K. Sengstock, and D. Pfannkuche *Phys. Rev. Lett.*, vol. 101, p. 050402, 2008.

[44] R. M. Lutchyn, S. Tewari, and S. D. Sarma *Physical Review* **A**, vol. 79, p. 011606(R), 2009.

[45] S. Tewari, R. M. Lutchyn, and S. D. Sarma *Phys. Rev.* **B**, vol. 80, p. 054511, 2009.

[46] H. Kamerlingh Onnes *Comm. Phys. ab. Univ. Leiden*, vol. 119, p. 120 122, 1911.

[47] P. Kapitsa *nature*, vol. 141, p. 74, 1938.

[48] J. F. Allen and A. D. Misener *nature*, vol. 141, p. 75, 1938.

[49] D. D. Osheroff, R. C. Richardson, and D. M. Lee *Phys. Rev. Lett.*, vol. 28, p. 885, 1972.

[50] F. London *Phys. Rev.*, vol. 54, p. 947, 1938.

[51] S. N. Bose *Z. Phys.*, vol. 26, p. 178, 1924.

[52] A. Einstein *Peuss. Akad. Wiss. Berlin Ber.*, vol. 22, p. 261, 1924.

[53] A. Einstein *Peuss. Akad. Wiss. Berlin Ber.*, vol. 23, p. 3, 1925.

[54] L. Landau *J. Phys. U. S. S. R.*, vol. 5, p. 71, 1941.

[55] L. Landau *J. Phys. U. S. S. R.*, vol. 11, p. 91, 1947.

[56] N. N. Bogoliubov *J. Phys. U. S. S. R.*, vol. 11, p. 23, 1947.

[57] L. N. Cooper *Phys. Rev.*, vol. 104, p. 1189, 1956.

[58] J. Bardeen, L. N. Cooper, and J. R. Schrieffer *Phys. Rev.*, vol. 106, p. 162, 1957.

[59] M. R. Schafroth *Phys. Rev.*, vol. 96, p. 1149, 1954.

[60] D. M. Eagles *Phys. Rev.*, vol. 186, p. 456, 1969.

[61] A. J. Leggett *J. Phys.* **C**, vol. 41, p. 7, 1980.

[62] P. Nozières and S. Schmitt-Rink *J. Low temp. Phys.*, vol. 59, p. 195, 1985.

[63] J. G. Bednorz and K. Mueller *Z. Physik* **B**, vol. 64, p. 189, 1986.

[64] M. H. Anderson, J. R. Ensher, M. R. Matthews, C. E. Wieman, and E. A. Cornell *Science*, vol. 269, p. 198, 1995.

[65] C. C. Bradley, C. A. Sackett, J. J. Tollett, and R. G. Hulet *Phys. Rev. Lett.*, vol. 75, p. 1687, 1995.

[66] K. B. Davis, M. O. Mewes, M. R. Andrews, N. J. van Druten, D. S. Durfee, D. M. Kurn, and W. Ketterle *Phys. Rev. Lett.*, vol. 75, p. 3969, 1995.

[67] D. S. Jin, . R. Ensher, M. R. Matthews, C. E. Wieman, and E. A. Cornell *Phys. Rev. Lett.*, vol. 77, p. 420, 1996.

[68] M. O. Mewes, M. R. Andrews, N. J. van Druten, D. M. Stamper-Kurn, D. S. Durfee, T. C. G., and W. Ketterle *Phys. Rev. Lett.*, vol. 77, p. 988, 1996.

[69] M. R. Andrews, T. C. G., H. J. Miesner, D. S. Durfee, D. M. Kurn, and W. Ketterle *Science*, vol. 275, p. 637, 1997.

[70] J. Stenger, S. Inouye, D. M. Stamper-Kurn, H. J. Miesner, A. P. Chikkatur, and W. Ketterle *Nature*, vol. 396, p. 345, 1998.

[71] F. S. Cataliotti, S. Burger, C. Fort, P. Maddaloni, F. Minardi, A. Trombetttoni, A. Smerzi, and M. Inguscio *Science*, vol. 293, p. 843, 2001.

[72] M. Albiez, R. Gati, J. Fölling, S. Hunsmann, M. Cristiani, and M. K. Oberthaler *Phys. Rev. Lett.*, vol. 95, p. 010402, 2005.

[73] M. Greiner, O. Mandel, T. Rom, A. Altmeyer, A. Widera, T. W. Hänsch, and I. Bloch *Physica* B, vol. 329, p. 11, 2003.

[74] S. Trotzky, P. Cheinet, S. Fölling, M. Feld, U. Schnorrberger, A. M. Rey, A. Polkovnikov, E. A. Demler, M. D. Lukin, and I. Bloch *Science*, vol. 319, p. 295, 2008.

[75] S. Fölling, S. Trotzky, P. Cheinet, M. Feld, R. Saers, A. Widera, T. Müller, and I. Bloch *Nature*, vol. 448, p. 1029, 2007.

[76] B. De Marco, C. Lannert, S. Vishveshwara, and T. C. Wei *Phys. Rev.* A, vol. 71, p. 063601, 2005.

[77] M. Schellekens, R. Hoppeler, A. Perrin, J. Viana Gomes, D. Boiron, A. Aspect, and C. I. Westbrook *Science*, vol. 310, p. 648, 2005.

[78] B. De Marco and D. C. . Jin *Science*, vol. 285, p. 1703, 1999.

[79] M. W. Zwierlein, A. Schirotzek, C. H. Schunck, and W. Ketterle *Science*, vol. 311, p. 492, 2006.

[80] M. W. Zwierlein, C. H. Schunck, A. Schirotzek, and W. Ketterle *Science*, vol. 442, p. 54, 2006.

[81] G. B. Partridge, W. Li, R. I. Kamar, Y. A. Liao, and R. G. Hulet *Science*, vol. 311, p. 503, 2006.

[82] Y. Shin, C. H. Schunck, A. Schirotzek, and W. Ketterle *Nature*, vol. 451, p. 689, 2008.

[83] G. Modugno, F. Ferlaino, R. Heidemann, G. Roati, and M. Inguscio *Phys. Rev.* **A**, vol. 68, p. 011601R, 2003.

[84] T. Stöferle, H. Moritz, K. Günter, M. Köhl, and T. Esslinger *Phys. Rev. Lett.*, vol. 96, p. 030401, 2006.

[85] J. K. Chin, D. E. Miller, Y. Liu, C. Stan, W. Setiawan, C. Sanner, K. Xu, and W. Ketterle *Nature*, vol. 443, p. 961, 2006.

[86] T. Rom, T. Best, D. van Oosten, U. Schneider, S. Fölling, B. Paredes, and I. Bloch *Nature*, vol. 444, p. 733, 2006.

[87] T. Taglieber, A. C. Voight, T. Aoki, T. W. Hänsch, and K. Dieckmann *Phys. Rev. Lett.*, vol. 100, p. 010401, 2008.

[88] E. Wille, F. M. Spiegelhalder, G. Kerner, D. Naik, A. Trenkwalder, G. Hendl, F. Schreck, R. Grimm, T. G. Tiecke, J. T. M. Walraven, S. J. J. M. F. Kokkelmans, E. Tiesinga, and P. S. Julienne *Phys. Rev. Lett.*, vol. 100, p. 053201, 2008.

[89] R. Jördens, N. Strohmaier, K. Günter, H. Moritz, and T. Esslinger *Nature*, vol. 455, p. 204, 2008.

[90] U. Schneider, L. Hackermüller, S. Will, B. Th., I. Bloch, T. A. Costi, R. W. Helmes, D. Rasch, and A. Rosch *Science*, vol. 322, p. 1520, 2008.

[91] S. Georgini, L. P. Pitaevskii, and S. Stingari *Rev. Mod. Phys.*, vol. 80, p. 1215, 2008, longer version in arXiv:0706. 3360.

[92] I. Bloch, J. Dailbard, and W. Zwerger *Rev. Mod. Phys.*, vol. 80, p. 885, 2008.

[93] I. Bloch and M. Greiner *Adv. At. Mol. Phys.*, vol. 52, p. 1, 2005.

[94] M. Lewenstein, A. Sanpera, V. Ahufinger, B. Damski, A. Sen, and U. Sen *Advances in Physics*, vol. 56, p. 243, 2007.

[95] W. Ketterle and M. W. Zwierlein *in Ultracold Fermi Gases, Proceedings of the International School of Physics "Enrico Fermi", Course CLXIV, Varenna, 20 - 30 June 2006; e-print: arXiv: 0801.2500*, 2006.

[96] W. Hofstetter *Forschung Frankfurt*, vol. 4/2006, p. 45, 2006.

[97] C. J. Pethick and H. Smith, *Bose-Einstein Condensation in Dilute Gases*. second edition, Cambridge university press, 2008.

[98] J. Reichel, F. Bardou, M. B. Dahan, E. Peik, S. Rand, C. Salomon, and C. Cohen-Tannoudji *Phys. Rev. Lett.*, vol. 75, p. 4575, 1995.

[99] H. F. Hess *Phys. Rev.* **B**, vol. 34, p. 3476, 1986.

[100] W. Ketterle and N. J. van Druten *Adv. At. Mol. Opt. Phys.*, vol. 37, p. 181, 1996.

[101] H. Metcalf and P. van der Straten, *Laser Cooling and Trapping*. Springer, New York, 1999.

[102] J. Jones *Proc. Roy. Soc.* **A**, vol. 106, p. 463, 1924.

[103] E. Tiesinga, B. J. Verhaar, and H. T. C. Stoof *Phys. Rev.* **A**, vol. 47, p. 4114, 1993.

[104] H. Feshbach *Ann. Phys.*, vol. 5, p. 337, 1958.

[105] U. Fano *Phys. Rev.*, vol. 124, p. 1866, 1961.

[106] P. Courteille, R. S. Freeland, D. J. Heinzen, F. A. van Abeelen, and B. J. Verhaar *Phys. Rev. Lett.*, vol. 81, p. 69, 1998.

[107] S. Inouye, K. B. Davis, M. R. Andrewa, J. Stenger, H.-J. Miesner, D. M. Stamper-Kurn, and W. Keterle *Nature*, vol. 392, p. 151, 1998.

[108] J. Stenger, S. Inouye, M. R. Andrewa, H.-J. Miesner, D. M. Stamper-Kurn, and W. Keterle *Phys. Rev. Lett.*, vol. 82, p. 2422, 1999.

[109] D. S. Petrov, C. Salomon, and G. V. Shlyapnikov *Phys. Rev. Lett.*, vol. 93, p. 090404, 2004.

[110] R. Combescot *Phys. Rev. Lett.*, vol. 91, p. 120401, 2003.

[111] S. Simonucci, P. Pieri, and G. C. Strinati *EPL*, vol. 69, p. 713, 2005.

[112] G. M. Bruun and C. Pethick *Phys. Rev. Lett.*, vol. 92, p. 140404, 2004.

[113] G. M. Bruun *Phys. Rev.* **A**, vol. 70, p. 053602, 2004.

[114] S. De Palo, M. L. Chiofalo, M. J. Holland, and S. J. J. M. F. Kokkelmans *Phys. Lett.* **A**, vol. 327, p. 490, 2004.

[115] T. Loftus, C. A. Regal, C. Ticknor, J. L. Bohn, and D. S. Jin *Phys. Rev. Lett.*, vol. 88, p. 173201, 2002.

[116] C. A. Regal, M. Greiner, and D. S. Jin vol. 92, p. 040403, 2004.

[117] M. Bartenstein, S. Altmeyer, S. Riedel, S. Jochim, R. Geursen, C. Chin, J. H. Denschlag, and R. Grimm *Phys. Rev. Lett.*, vol. 92, p. 120401, 2004.

[118] M. W. Zwierlein, C. A. Stan, C. H. Schunck, S. M. F. Raupach, A. J. Kerman, and W. Ketterle *Phys. Rev. Lett.*, vol. 92, p. 120403, 2004.

[119] S. Jochim, M. Bartenstein, A. Altmeyer, G. Hendl, C. Chin, J. H. Denschlag, and R. Grimm *Science*, vol. 302, p. 2101, 2003.

[120] G. B. Partridge, K. E. Strecker, R. I. Kamer, M. W. Jack, and R. G. Hulet *Phys. Rev. Lett.*, vol. 95, p. 020404, 2005.

[121] K. R. Strecker, G. B. Partridge, and H. R. G. *Phys. Rev. Lett.*, vol. 91, p. 080406, 2003.

[122] E. Timmermans, K. Furuya, M. P. W., and A. K. Kerman *Phys. Lett.* **A**, vol. 285, p. 228, 2001.

[123] R. A. Duine and H. T. C. Stoff *Phys. Rep.*, vol. 396, p. 115, 2004.

BIBLIOGRAPHY 139

[124] T. Köhler, K. Goral, and P. S. Julienne *Rev. Mod. Phys.*, vol. 78, p. 1311, 2006.

[125] F. Ferlaino, S. Knoop, and R. Grimm *Cheptar of book: "Cold Molecules: Theory, Experiment, Applications" edited by R. V. Krems, B. Friedrich and W. C. Stwalley; arXiv:0809.3920v1*, 2009.

[126] M. Greiner, C. A. Regel, and . D. C. Jin *Nature*, vol. 426, p. 537, 2003.

[127] M. W. Zwierlein, C. A. Stan, C. H. Schunk, S. M. F. Raupach, S. Gupta, Z. Hadzibabic, and W. Keterle *Phys. Rev. Lett.*, vol. 91, p. 250401, 2003.

[128] T. Bourdel, L. Khaykovich, J. Cubizolles, J. Zhang, F. Chevy, M. Teichmann, L. Tarruell, S. J. J. M. F. Kokkelmans, and C. Salomon *Phys. Rev. Lett.*, vol. 93, p. 050401, 2004.

[129] J. Cubizolles, T. Bourdel, S. J. J. M. F. Kokkelmans, G. V. Shlyapnikov, and C. Salomon *Phys. Rev. Lett.*, vol. 91, p. 240401, 2003.

[130] C. A. R. Sá de Melo *Physics Today*, vol. 61, p. 45, 2008.

[131] A. Privitera, *Lattice Approach to the BCS-BEC Crossover in Dilute Systems: a MF and DMFT Approach*. PhD thesis, Sapienza, Università di Roma, 2008.

[132] T. L. Ho *Phys. Rev. Lett.*, vol. 92, p. 090402, 2004.

[133] Ákos Rapp, G. Zaránd, C. Honerkamp, and W. Hofstetter *Phys. Rev. Lett.*, vol. 98, p. 160405, 2007.

[134] Ákos Rapp, W. Hofstetter, and G. Zaránd *Phys. Rev. B*, vol. 77, p. 144520, 2008.

[135] A. J. Daley, J. M. Taylor, S. Diehl, M. Baranov, and P. Zoller *Phys. Rev. Lett.*, vol. 102, p. 040402, 2009.

[136] T. B. Ottenstein, T. Lompe, M. Kohnen, A. N. Wenz, and S. Jochim *Phys. Rev. Lett.*, vol. 101, p. 203202, 2008.

[137] C. Cohen-Tannoudji, J. Dupon-Roc, and G. Grynberg, *Atom-Photon Interactions*. Wiley-VCH, Berlin, 1992.

[138] S. Chu, J. E. Bjorkholm, A. Ashkin, and A. Cable *Phys. Rev. Lett.*, vol. 57, p. 314, 1986.

[139] R. Grimm, M. Weidemüller, and Y. Ovchinnikov *Adv. At. Mol. Opt.*, vol. 42, p. 95, 2000.

[140] K. I. Petsas, A. B. Coates, and G. Grynberg *Phys. Rev.* **A**, vol. 50, p. 5173, 1994.

[141] Z. Hadzibabic, S. Stock, B. Battelier, V. Bretin, and J. Dalibard *Phys. Rev. Lett.*, vol. 93, p. 180403, 2004.

[142] S. Peil, J. V. Porto, B. L. Tolra, J. M. Obrecht, B. E. King, M. Subbotin, S. L. Rolston, and W. D. Phillips *Phys. Rev.* **A**, vol. 67, p. 051603, 2003.

[143] W. Zwerger *J. Opt.* **B**, vol. 5, p. S9, 2003.

[144] F. F. Assaad, R. Preuss, A. Muramatsu, and W. Hanke *Jour. of Low Temp. Physics*, vol. 95, p. 251, 1994.

[145] R. Staudt, M. Dzierzawa, and A. Muramatsu *Eur. Phys. J.* **B**, vol. 17, p. 411, 2000.

[146] M. Rigol and A. Muramatsu *Phys. Rev.* **A**, vol. 69, p. 053612, 2004.

[147] W. Metzner and D. Vollhardt *Phys. Rev. Lett.*, vol. 62, p. 324, 1989.

[148] A. Georges, G. Kotliar, W. Krauth, and J. Rozenberg *Rev. Mod. Phys.*, vol. 68, p. 13, 1996.

[149] A. Georges *Lectures on the Physics of Highly Correlated Electron Systems VIII; arXiv:cond-mat/0403123v1*, 2004.

[150] E. Müller-Hartmann *Zeitschrift für Physik* **B**: *Condensed Matter.*, vol. 74, p. 507, 1989.

[151] E. Müller-Hartmann *Zeitschrift für Physik* **B**: *Condensed Matter.*, vol. 76, p. 211, 1989.

[152] E. Müller-Hartmann *International Journal of Modern Physics* **B**, vol. 3, p. 2169, 1989.

[153] M. Jarrell *Phys. Rev. Lett.*, vol. 69, p. 168, 1992.

[154] A. Georges and G. Kotliar *Phys. Rev.* **B**, vol. 45, p. 6479, 1992.

[155] A. Georges, G. Kotliar, and W. Krauth *Z. Phys.* **B**, vol. 92, p. 313, 1993.

[156] P. Limelette, P. Wzietek, S. Florens, A. Georges, T. A. Costi, C. Pasquier, D. Jerome, C. Meziere, and B. P. *Phys. Rev. Lett*, vol. 91, p. 016401, 2003.

[157] J. Hubbard *Proc. Roy. Soc. A*, vol. 281, p. 401, 1964.

[158] D. S. Rokhsar and B. G. Kotliar *Phys. Rev.* **B**, vol. 44, p. 10328, 1991.

[159] K. Sheshadri, H. R. Krishnamurthy, R. Pandit, and T. V. Ramakrishnan *Europhysics Letters*, vol. 22, p. 257, 1993.

[160] D. van Oosten, P. van der Straten, and H. T. C. Stoof *Phys. Rev.* **A**, vol. 63, p. 053601, 2001.

[161] K. Byczuk and D. Vollhardt *Phys. Rev.* **B**., vol. 77, p. 235106, 2008.

[162] A. Hubener, M. Snoek, and W. Hofstetter *Phys. Rev.* **B**, vol. 80, p. 245109, 2008.

[163] M. Snoek, I. Titvinidze, C. Tőke, K. Byczuk, and W. Hofstetter *New Journal of Physics*, vol. 8, p. 093008, 2008.

[164] M. Potthoff and W. Nolting *Phys. Rev.* **B**, vol. 59, p. 2549, 1999.

[165] M.-T. Tran *Phys. Rev.* **B**, vol. 73, p. 205110, 2006.

[166] M.-T. Tran *Phys. Rev.* **B**, vol. 76, p. 245122, 2007.

[167] Y. Song, R. Wortis, and W. A. Atkinson *Phys. Rev.* **B**, vol. 77, p. 054202, 2008.

[168] R Helmes, T. A. Costi, and A. Rosch *Phys. Rev. Lett*, vol. 100, p. 056403, 2008.

[169] R. Helmes, *Dynamical Mean Field Theory of inhomogeneous correlated systems*. PhD thesis, Universität zu Köln, 2008.

[170] K. G. Wilson *Rev. Mod. Phys.*, vol. 47, p. 773, 1975.

[171] H. R. Krishna-murthy, K. G. Wilson, and J. W. Wilkins *Phys. Rev. Lett.*, vol. 35, p. 1101, 1975.

[172] H. R. Krishna-murthy, J. W. Wilkins, and K. G. Wilson *Phys. Rev.* **B**, vol. 21, p. 1003, 1980.

[173] H. R. Krishna-murthy, J. W. Wilkins, and K. G. Wilson *Phys. Rev.* **B**, vol. 21, p. 1044, 1980.

[174] W. Hofstetter *Phys. Rev. Lett.*, vol. 85, p. 1508, 2000.

[175] R. Zitzler, T. Pruschke, and R. Bulla *Eur. Phys. J.* **B**, vol. 27, p. 473, 2003.

[176] R. Bulla, A. C. Hewson, and T. Pruschke *J. Phys: Condens. matter*, vol. 10, p. 8365, 1998.

[177] R. Bulla, T. A. Costi, and T. Pruschke *Rev. Mod. Phys.*, vol. 80, p. 395, 2008.

[178] W. Hofstetter, *Renormalization Group methods for Quantum Impurity Systems*. PhD thesis, Universität Augsburg, 2000.

[179] M. Caffarel and W. Krauth *Phys. Rev. Lett.*, vol. 72, p. 1545, 1994.

[180] Q. Si, M. J. Rozenberg, K. Kotliar, and A. E. Ruckenstein *Phys. Rev. Lett.*, vol. 72, p. 2761, 1994.

[181] A. Toschi, M. Capone, and C. Castellani *Phys. Rev.* **B.**, vol. 72, p. 235118, 2005.

[182] J. E. Hirsch and R. M. Fye *Phys. Rev. Lett*, vol. 56, p. 2521, 1986.

[183] N. Blümer *Preprint: arXiv:0801. 1222*, 2008.

[184] J. Joo and V. Oudovenko *Phys. Rev.* **B**, vol. 64, p. 193102, 2001.

[185] K. Aryanpour, W. E. Pickett, and R. T. Scalettar *Phys. Rev.* **B**, vol. 74, p. 085117, 2006.

[186] P. W. Anderson *J. Phys.* **C**, vol. 3, p. 2436, 1970.

[187] J. R. Schrieffer and W. P. A. *Phys. Rev.*, vol. 149, p. 491, 1966.

[188] C. Lanczos *J. Res. Natl. Bur. Stand.*, vol. 45, p. 255, 1950.

[189] F. Gerbier, A. Widera, S. Fölling, O. Mandel, T. Gericke, and I. Bloch *Phys. Rev. Lett.*, vol. 95, p. 050404, 2005.

[190] F. Gerbier, A. Widera, S. Fölling, O. Mandel, T. Gericke, and I. Bloch *Phys. Rev.* **A**, vol. 72, p. 053606, 2005.

[191] V. G. Rousseau, D. P. Arovas, M. Rigol, F. Hébert, G. G. Batrouni, and R. T. Scalettar *Phys. Rev.* **B**, vol. 73, p. 174516, 2006.

[192] P. Buensante and A. Vezzani *Phys. Rev. A*, vol. 70, p. 030608, 2004.

[193] P. Buensante, P. Penna, and A. Vezzani *Laser. Phys.*, vol. 15, p. 361, 2005.

[194] B. M. Andersen and G. M. Bruun *Phys. Rev. A*, vol. 76, p. 041602, 2007.

[195] B. I. Shraiman and E. D. Siggia *Phys. Rev. Lett.*, vol. 62, p. 1564, 1989.

[196] H. J. Schulz *Phys. Rev. Lett.*, vol. 64, p. 1445, 1990.

[197] J. K. Freericks and M. Jarrell *Phys. Rev. Lett.*, vol. 74, p. 186, 1995.

[198] T. Pruschke and R. Zitzler *J. Phys. : Condens. Matter*, vol. 15, p. 7867, 2003.

[199] K. Byczuk *Condensed Matter Physics in the Prime of the 21st Century: Phenomena, Materials, Ideas, Methods ed J Jedrzejewski (Singapore: World Scientific)*, pp. 1–33, 2008.

[200] M. Holland, S. J. J. M. F. Kokkelmans, M. L. Chiofalo, and R. Walser *Phys. Rev. Lett.*, vol. 87, p. 120406, 2001.

[201] L. D. Carr and M. J. Holland *Phys. Rev. A*, vol. 72, p. 031604R, 2005.

[202] A. Koetsier, D. B. M. Dickerscheid, and H. T. C. Stoof *Phys. Rev. A*, vol. 74, p. 033621, 2006.

[203] T. Busch, B.-G. Englert, K. Rzążewski, and M. Wilkens *Found. Phys.*, vol. 28, p. 549, 1998.

[204] D. S. Petrov, C. Salomon, and G. V. Shlyapnikov *Phys. Rev. A*, vol. 71, p. 012708, 2005.

[205] M. Abramowitz and e. I.A. Stegung, *Handbook of Mathematicsal Functions*. Dover, New York, 1972.

[206] W. Nolting, *Grundkurs: Theoretische Physik 7: Viel Teilchen-Theorie*. Verlag Zimmermann-Neufang, 1992.

Acknowledgments

First of all I would like to thank my supervisor Prof. Dr. *Walter Hofstetter* for giving me the opportunity to write my PHD thesis in his group. I am thankful to him for numerous helpfull discussions and very careful proofreading of this thesis.

I am grateful to Dr. *Michiel Snoek*, who was my collaborator of all works presented in this thesis, for a lot of helpful discussions as well as for very careful proofreading of this thesis and giving helpful advices.

I would like to thank *David Roosen*, with whom I was sharing office in Aachen as well as in Frankfurt. He helped me when we move from Aachen to Frankfurt and always was helping me with computer problems.

I am thankful to *Denis Semmler* for translating "Zussamenfassung" in German and *Ulf Bissbort* for helping me to make some figures using illustrator.

Special thanks to our group secretary *Daniela Wirth-Pagano*, not only for the constant and important support, but also for a kindness far beyond her working duties.

I thank to all present and former members of our group *David Roosen, Ulf Bissbort, Denis Semmler*, Dr. *Michiel Snoek, Falk May, Julia Wernsdorfer, Andreas Hubener, Johannes Ferber*, Dr. *Antonio Privitera*, Dr. *Reza Bakhtiari*, Dr. *Csaba Tőke* and *Yongqiang Li* for making our group a nice place to work and for the good times which we spent together.

I would like to thank my former supervisor Prof. Dr. *George Japaridze* who opened me the door to science.

I also thank my old friends Dr. *Paata Kakashvili, Michael Sekania*, Prof. Dr. *Temo Vekua* for constantly keeping in touch with me and all the fun which we had together in Georgia, Germany, Sweden, Denmark, France and Italy.

My warm greetings to my brother Dr. *George Titvinidze*. For support and a lot of fun which we have together in Georgia and Germany.

Finally I would like to thank my mother *Tsisana Imnadze*, and aunts *Meri* and *Nino Titvinidze* for endless support and great love.

For all gifts of health, strength and mind I bring my thanksgivings to God.

I want morebooks!

Buy your books fast and straightforward online - at one of world's fastest growing online book stores! Environmentally sound due to Print-on-Demand technologies.

Buy your books online at
www.morebooks.shop

Kaufen Sie Ihre Bücher schnell und unkompliziert online – auf einer der am schnellsten wachsenden Buchhandelsplattformen weltweit! Dank Print-On-Demand umwelt- und ressourcenschonend produziert.

Bücher schneller online kaufen
www.morebooks.shop

KS OmniScriptum Publishing
Brivibas gatve 197
LV-1039 Riga, Latvia
Telefax: +371 686 204 55

info@omniscriptum.com
www.omniscriptum.com

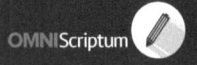

Printed by Books on Demand GmbH, Norderstedt / Germany